ORDER OUT OF CHAOS

ORDER OUT OF CHAOS

MAN'S NEW DIALOGUE WITH NATURE

Ilya Prigogine
and
Isabelle Stengers

Foreword by
Alvin Toffler

BANTAM BOOKS
TORONTO · NEW YORK · LONDON · SYDNEY

ORDER OUT OF CHAOS:
MAN'S NEW DIALOGUE WITH NATURE

A Bantam Book / April 1984
4 printings through October 1988

Cover: One-, two-, three-, and four-armed vortices in an active excitable chemical medium. Courtesy of V.I. Krinsky, Institute of Biophysics, USSR Academy of Sciences.

Book design by Barbara N. Cohen

Library of Congress Cataloging in Publication Data

Prigogine, I. (Ilya)
Order out of chaos.

Based on the authors' la nouvelle alliance.
Includes bibliographical references and index.
1. Science—Philosophy. 2. Physics—Philosophy.
3. Thermodynamics. 4. Irreversible processes.
I. Stengers, Isabelle. II. Prigogine, I. (Ilya)
La nouvelle alliance. III. Title.
Q175.P8823 1984 501 83-21403
ISBN 0-553-34082-4

Published simultaneously in the United States and Canada

Bantam Books are published by Bantam Books, a division of Bantam Doubleday Dell Publishing Group, Inc. Its trademark, consisting of the words "Bantam Books" and the portrayal of a rooster, is Registered in U.S. Patent and Trademark Office and in other countries. Marca Registrada. Bantam Books, 666 Fifth Avenue, New York, New York 10103.

PRINTED IN THE UNITED STATES OF AMERICA

FG 13 12 11 10 9 8 7 6 5 4

This book is dedicated to the memory of

Erich Jantsch
Aharon Katchalsky
Pierre Résibois
Léon Rosenfeld

TABLE OF CONTENTS

FOREWORD

SCIENCE AND CHANGE

by Alvin Toffler

One of the most highly developed skills in contemporary Western civilization is dissection: the split-up of problems into their smallest possible components. We are good at it. So good, we often forget to put the pieces back together again.

This skill is perhaps most finely honed in science. There we not only routinely break problems down into bite-sized chunks and mini-chunks, we then very often isolate each one from its environment by means of a useful trick. We say *ceteris paribus*—all other things being equal. In this way we can ignore the complex interactions between our problem and the rest of the universe.

Ilya Prigogine, who won the Nobel Prize in 1977 for his work on the thermodynamics of nonequilibrium systems, is not satisfied, however, with merely taking things apart. He has spent the better part of a lifetime trying to "put the pieces back together again"—the pieces in this case being biology and physics, necessity and chance, science and humanity.

Born in Russia in 1917 and raised in Belgium since the age of ten, Prigogine is a compact man with gray hair, cleanly chiseled features, and a laserlike intensity. Deeply interested in archaeology, art, and history, he brings to science a remarkable polymathic mind. He lives with his engineer-wife, Marina, and his son, Pascal, in Brussels, where a cross-disciplinary team is busy exploring the implications of his ideas in fields as disparate as the social behavior of ant colonies, diffusion reactions in chemical systems and dissipative processes in quantum field theory.

He spends part of each year at the Ilya Prigogine Center for Statistical Mechanics and Thermodynamics of the University of Texas in Austin. To his evident delight and surprise, he was awarded the Nobel Prize for his work on "dissipative structures" arising out of nonlinear processes in nonequilibrium systems. The coauthor of this volume, Isabelle Stengers, is a philosopher, chemist, and historian of science who served for a time as part of Prigogine's Brussels team. She now lives in Paris and is associated with the Musée de la Villette.

In *Order Out of Chaos* they have given us a landmark—a work that is contentious and mind-energizing, a book filled with flashing insights that subvert many of our most basic assumptions and suggest fresh ways to think about them.

Under the title *La nouvelle alliance,* its appearance in France in 1979 triggered a marvelous scientific free-for-all among prestigious intellectuals in fields as diverse as entomology and literary criticism.

It is a measure of America's insularity and cultural arrogance that this book, which is either published or about to be published in twelve languages, has taken so long to cross the Atlantic. The delay carries with it a silver lining, however, in that this edition includes Prigogine's newest findings, particularly with respect to the Second Law of thermodynamics, which he sets into a fresh perspective.

For all these reasons, *Order Out of Chaos* is more than just another book: It is a lever for changing science itself, for compelling us to reexamine its goals, its methods, its epistemology—its world view. Indeed, this book can serve as a symbol of today's historic transformation in science—one that no informed person can afford to ignore.

Some scholars picture science as driven by its own internal logic, developing according to its own laws in splendid isolation from the world around it. Yet many scientific hypotheses, theories, metaphors, and models (not to mention the choices made by scientists either to study or to ignore various problems) are shaped by economic, cultural, and political forces operating outside the laboratory.

I do not mean to suggest too neat a parallel between the nature of society and the reigning scientific world view or "paradigm." Still less would I relegate science to some "superstructure" mounted atop a socioeconomic "base," as Marxists are wont to do. But science is not an "independent variable." It is an open system embedded in society and linked to it by very dense feedback loops. It is powerfully influenced by its external environment, and, in a general way, its development is shaped by cultural receptivity to its dominant ideas.

Take that body of ideas that came together in the seventeenth and eighteenth centuries under the heading of "classical science" or "Newtonianism." They pictured a world in which every event was determined by initial conditions that were, at least in principle, determinable with precision. It was a world in which chance played no part, in which all the pieces came together like cogs in a cosmic machine.

The acceptance of this mechanistic view coincided with the rise of a factory civilization. And divine dice-shooting seems hardly enough to account for the fact that the Age of the Machine enthusiastically embraced scientific theories that pictured the entire universe as a machine.

This view of the world led Laplace to his famous claim that, given enough facts, we could not merely predict the future but retrodict the past. And this image of a simple, uniform, mechanical universe not only shaped the development of science, it also spilled over into many other fields. It influenced the framers of the American Constitution to create a machine for governing, its checks and balances clicking like parts of a clock. Metternich, when he rode forth to create his balance of power in Europe, carried a copy of Laplace's writings in his baggage. And the dramatic spread of factory civilization, with its vast clanking machines, its heroic engineering breakthroughs, the rise of the railroad, and new industries such as steel, textile, and auto, seemed merely to confirm the image of the universe as an engineer's Tinkertoy.

Today, however, the Age of the Machine is screeching to a halt, if ages can screech—and ours certainly seems to. And the decline of the industrial age forces us to confront the painful limitations of the machine model of reality.

Of course, most of these limitations are not freshly discovered. The notion that the world is a clockwork, the planets timelessly orbiting, all systems operating deterministically in equilibrium, all subject to universal laws that an outside observer could discover—this model has come under withering fire ever since it first arose.

In the early nineteenth century, thermodynamics challenged the timelessness implied in the mechanistic image of the universe. If the world was a big machine, the thermodynamicists declared, it was running down, its useful energy leaking out. It could not go on forever, and time, therefore, took on a new meaning. Darwin's followers soon introduced a contradictory thought: The world-machine might be running down, losing energy and organization, but biological systems, at least, were running *up,* becoming more, not less, organized.

By the early twentieth century, Einstein had come along to put the observer back into the system: The machine looked different—indeed, for all practical purposes it *was* different—depending upon where you stood within it. But it was still a deterministic machine, and God did not throw dice. Next, the quantum people and the uncertainty folks attacked the model with pickaxes, sledgehammers, and sticks of dynamite.

Nevertheless, despite all the ifs, ands, and buts, it remains fair to say, as Prigogine and Stengers do, that the machine paradigm is still the "reference point" for physics and the core model of science in general. Indeed, so powerful is its continuing influence that much of social science, and especially economics, remains under its spell.

The importance of this book is not simply that it uses original arguments to challenge the Newtonian model, but also that it shows how the still valid, though much limited, claims of Newtonianism might fit compatibly into a larger scientific image of reality. It argues that the old "universal laws" are not universal at all, but apply only to local regions of reality. And these happen to be the regions to which science has devoted the most effort.

Thus, in broad-stroke terms, Prigogine and Stengers argue that traditional science in the Age of the Machine tended to emphasize stability, order, uniformity, and equilibrium. It concerned itself mostly with closed systems and linear relationships in which small inputs uniformly yield small results.

With the transition from an industrial society based on heavy inputs of energy, capital, and labor to a high-technology society in which information and innovation are the critical resources, it is not surprising that new scientific world models should appear.

What makes the Prigoginian paradigm especially interesting is that it shifts attention to those aspects of reality that characterize today's accelerated social change: disorder, instability, diversity, disequilibrium, nonlinear relationships (in which small inputs can trigger massive consequences), and temporality—a heightened sensitivity to the flows of time.

The work of Ilya Prigogine and his colleagues in the so-called "Brussels school" may well represent the next revolution in science as it enters into a new dialogue not merely with nature, but with society itself.

The ideas of the Brussels school, based heavily on Prigogine's work, add up to a novel, comprehensive theory of change.

Summed up and simplified, they hold that while some parts of the universe may operate like machines, these are closed systems, and closed systems, at best, form only a small part of the physical universe. Most phenomena of interest to us are, in fact, *open* systems, exchanging energy or matter (and, one might add, information) with their environment. Surely biological and social systems are open, which means that the attempt to understand them in mechanistic terms is doomed to failure.

This suggests, moreover, that most of reality, instead of being orderly, stable, and equilibrial, is seething and bubbling with change, disorder, and process.

In Prigoginian terms, all systems contain subsystems, which are continually "fluctuating." At times, a single fluctuation or a combination of them may become so powerful, as a result of positive feedback, that it shatters the preexisting organization. At this revolutionary moment—the authors call it a "singular moment" or a "bifurcation point"—it is inherently impossible to determine in advance which direction change will take: whether the system will disintegrate into "chaos" or leap to a new, more differentiated, higher level of "order" or organization, which they call a "dissipative structure." (Such physical or chemical structures are termed dissipative because, compared with the simpler structures they replace, they require more energy to sustain them.)

One of the key controversies surrounding this concept has to do with Prigogine's insistence that order and organization can actually arise "spontaneously" out of disorder and chaos through a process of "self-organization."

To grasp this extremely powerful idea, we first need to make a distinction between systems that are in "equilibrium," systems that are "near equilibrium," and systems that are "far from equilibrium."

Imagine a primitive tribe. If its birthrate and death rate are equal, the size of the population remains stable. Assuming adequate food and other resources, the tribe forms part of a local system in ecological equilibrium.

Now increase the birthrate. A few additional births (without an equivalent number of deaths) might have little effect. The system may move to a near-equilibrial state. Nothing much happens. It takes a big jolt to produce big consequences in systems that are in equilibrial or near-equilibrial states.

But if the birthrate should suddenly soar, the system is pushed into a far-from-equilibrium condition, and here nonlinear relationships prevail. In this state, systems do strange things. They become inordinately sensitive to external influences. Small inputs yield huge, startling effects. The entire system may reorganize itself in ways that strike us as bizarre.

Examples of such self-reorganization abound in *Order Out of Chaos*. Heat moving evenly through a liquid suddenly, at a certain threshold, converts into a convection current that radically reorganizes the liquid, and millions of molecules, as if on cue, suddenly form themselves into hexagonal cells.

Even more spectacular are the "chemical clocks" described by Prigogine and Stengers. Imagine a million white ping-pong balls mixed at random with a million black ones, bouncing around chaotically in a tank with a glass window in it. Most of the time, the mass seen through the window would appear to be gray, but now and then, at irregular moments, the sample seen through the glass might seem black or white, depending on the distribution of the balls at that moment in the vicinity of the window.

Now imagine that suddenly the window goes all white, then all black, then all white again, and on and on, changing its color completely at fixed intervals—like a clock ticking.

Why do all the white balls and all the black ones suddenly organize themselves to change color in time with one another?

By all the traditional rules, this should not happen at all. Yet, if we leave ping-pong behind and look at molecules in certain chemical reactions, we find that precisely such a self-organization or ordering can and does occur—despite what classical physics and the probability theories of Boltzmann tell us.

In far-from-equilibrium situations other seemingly spontaneous, often dramatic reorganizations of matter within time and space also take place. And if we begin thinking in terms of two or three dimensions, the number and variety of such possible structures become very great.

Now add to this an additional discovery. Imagine a situation in which a chemical or other reaction produces an enzyme whose presence then encourages further production of the same enzyme. This is an example of what computer scientists would call a positive-feedback loop. In chemistry it is called "auto-catalysis." Such situations are rare in inorganic chemistry. But in recent decades the molecular biologists have found that such loops (along with inhibitory or "negative" feedback and more complicated "cross-catalytic" processes) are the very stuff of life itself. Such processes help explain how we go from little lumps of DNA to complex living organisms.

More generally, therefore, in far-from-equilibrium conditions we find that very small perturbations or fluctuations can become amplified into gigantic, structure-breaking waves. And this sheds light on all sorts of "qualitative" or "revolutionary" change processes. When one combines the new insights gained from studying far-from-equilibrium states and nonlinear processes, along with these complicated feedback systems, a whole new approach is opened that makes it possible to relate the so-called hard sciences to the softer sciences of life—and perhaps even to social processes as well.

(Such findings have at least analogical significance for social, economic or political realities. Words like "revolution," "economic crash," "technological upheaval," and "paradigm shift" all take on new shades of meaning when we begin thinking of them in terms of fluctuations, feedback amplification, dissipative structures, bifurcations, and the rest of the Prigoginian conceptual vocabulary.) It is these panoramic vistas that are opened to us by *Order Out of Chaos*.

Beyond this, there is the even more puzzling, pervasive issue of time.

Part of today's vast revolution in both science and culture is a reconsideration of time, and it is important enough to merit a brief digression here before returning to Prigogine's role in it.

Take history, for example. One of the great contributions to historiography has been Braudel's division of time into three scales—"geographical time," in which events occur over the course of aeons; the much shorter "social time" scale by which economies, states, and civilizations are measured; and the even shorter scale of "individual time"—the history of human events.

In social science, time remains a largely unmapped terrain. Anthropology has taught us that cultures differ sharply in the way they conceive of time. For some, time is cyclical—history endlessly recurrent. For other cultures, our own included, time is a highway stretched between past and future, and people or whole societies march along it. In still other cultures, human lives are seen as stationary in time; the future advances toward us, instead of us toward it.

Each society, as I've written elsewhere, betrays its own characteristic "time bias"—the degree to which it places emphasis on past, present, or future. One lives in the past. Another may be obsessed with the future.

Moreover, each culture and each person tends to think in terms of "time horizons." Some of us think only of the immediate—the now. Politicians, for example, are often criticized for seeking only immediate, short-term results. Their time horizon is said to be influenced by the date of the next election. Others among us plan for the long term. These differing time horizons are an overlooked source of social and political friction—perhaps among the most important.

But despite the growing recognition that cultural conceptions of time differ, the social sciences have developed little in the way of a coherent theory of time. Such a theory might reach across many disciplines, from politics to group dynamics and interpersonal psychology. It might, for example, take account of what, in *Future Shock,* I called "durational expectancies"—our culturally induced assumptions about how long certain processes are supposed to take.

We learn very early, for example, that brushing one's teeth should last only a few minutes, not an entire morning, or that when Daddy leaves for work, he is likely to be gone approximately eight hours, or that a "mealtime" may last a few minutes or hours, but never a year. (Television, with its division of the day into fixed thirty- or sixty-minute intervals, subtly shapes our notions of duration. Thus we normally expect the hero in a melodrama to get the girl or find the money or win the war in the last five minutes. In the United States we expect commercials to break in at certain intervals.) Our minds are filled with such durational assumptions. Those of children are much different from those of fully socialized adults, and here again the differences are a source of conflict.

Moreover, children in an industrial society are "time trained"—they learn to read the clock, and they learn to distinguish even quite small slices of time, as when their parents tell them, "You've only got three more minutes till bedtime!" These sharply honed temporal skills are often absent in slower-moving agrarian societies that require less precision in daily scheduling than our time-obsessed society.

Such concepts, which fit within the social and individual time scales of Braudel, have never been systematically developed in the social sciences. Nor have they, in any significant way, been articulated with our scientific theories of time, even though they are necessarily connected with our assumptions about physical reality. And this brings us back to Prigogine, who has been fascinated by the concept of time since boyhood. He once said to me that, as a young student, he was struck by a grand contradiction in the way science viewed time, and this contradiction has been the source of his life's work ever since.

In the world model constructed by Newton and his followers, time was an afterthought. A moment, whether in the present, past, or future, was assumed to be exactly like any other moment. The endless cycling of the planets—indeed, the operations of a clock or a simple machine—can, in principle, go either backward or forward in time without altering the basics of the system. For this reason, scientists refer to time in Newtonian systems as "reversible."

In the nineteenth century, however, as the main focus of physics shifted from dynamics to thermodynamics and the Second Law of thermodynamics was proclaimed, time suddenly became a central concern. For, according to the Second Law, there is an inescapable loss of energy in the universe. And, if the world machine is really running down and approaching the heat death, then it follows that one moment is no longer exactly like the last. You cannot run the universe backward to make up for entropy. Events over the long term cannot replay themselves. And this means that there is a directionality or, as Eddington later called it, an "arrow" in time.

The whole universe is, in fact, aging. And, in turn, if this is true, time is a one-way street. It is no longer reversible, but irreversible.

In short, with the rise of thermodynamics, science split down the middle with respect to time. Worse yet, even those who saw time as irreversible soon also split into two camps. After all, as energy leaked out of the system, its ability to sustain organized structures weakened, and these, in turn, broke down into less organized, hence more random elements. But it is precisely organization that gives any system internal diversity. Hence, as entropy drained the system of energy, it also reduced the differences in it. Thus the Second Law pointed toward an increasingly homogeneous—and, from the human point of view, pessimistic—future.

Imagine the problems introduced by Darwin and his followers! For evolution, far from pointing toward reduced organization and diversity, points in the opposite direction. Evolution proceeds from simple to complex, from "lower" to "higher" forms of life, from undifferentiated to differentiated structures. And, from a human point of view, all this is quite optimistic. The universe gets "better" organized as it ages, continually advancing to a higher level as time sweeps by.

In this sense, scientific views of time may be summed up as a contradiction within a contradiction.

It is these paradoxes that Prigogine and Stengers set out to illuminate, asking, "What is the specific structure of dynamic systems which permits them to 'distinguish' between past and future? What is the minimum complexity involved?"

The answer, for them, is that time makes its appearance with randomness: "Only when a system behaves in a sufficiently random way may the difference between past and future, and therefore irreversibility, enter its description."

In classical or mechanistic science, events begin with "initial conditions," and their atoms or particles follow "world lines" or trajectories. These can be traced either backward into the past or forward into the future. This is just the opposite of certain chemical reactions, for example, in which two liquids poured into the same pot diffuse until the mixture is uniform or homogeneous. These liquids do not de-diffuse themselves. At each moment of time the mixture is different, the entire process is "time-oriented."

For classical science, at least in its early stages, such processes were regarded as anomalies, peculiarities that arose from highly unlikely initial conditions.

It is Prigogine and Stengers' thesis that such time-dependent, one-way processes are not merely aberrations or deviations from a world in which time is reversible. If anything, the opposite might be true, and it is reversible time, associated with "closed systems" (if such, indeed, exist in reality), that may well be the rare or aberrant phenomenon.

What is more, irreversible processes are the source of order—hence the title *Order Out of Chaos*. It is the processes associated with randomness, openness, that lead to higher levels of organization, such as dissipative structures.

Indeed, one of the key themes of this book is its striking reinterpretation of the Second Law of thermodynamics. For according to the authors, entropy is not merely a downward slide toward disorganization. Under certain conditions, entropy itself becomes the progenitor of order.

What the authors are proposing, therefore, is a vast synthesis that embraces both reversible *and* irreversible time, and shows how they relate to one another, not merely at the level of macroscopic phenomena, but at the most minute level as well.

It is a breathtaking attempt at "putting the pieces back together again." The argument is complex, and at times beyond easy reach of the lay reader. But it flashes with fresh insight and suggests a coherent way to relate seemingly unconnected—even contradictory—philosophical concepts.

Here we begin to glimpse, in full richness, the monumental synthesis proposed in these pages. By insisting that irreversible time is not a mere aberration, but a characteristic of much of the universe, they subvert classical dynamics. For Prigogine and Stengers, it is not a case of either/or. Of course, reversibility still applies (at least for sufficiently long times)—but in closed systems only. Irreversibility applies to the rest of the universe.

Prigogine and Stengers also undermine conventional views of thermodynamics by showing that, under nonequilibrium conditions, at least, entropy may produce, rather than degrade, order, organization—and therefore life.

If this is so, then entropy, too, loses its either/or character. While certain systems run down, other systems simultaneously evolve and grow more coherent. This mutualistic, nonexclusive view makes it possible for biology and physics to coexist rather than merely contradict one another.

Finally, yet another profound synthesis is implied—a new relationship between chance and necessity.

The role of happenstance in the affairs of the universe has been debated, no doubt, since the first Paleolithic warrior accidently tripped over a rock. In the Old Testament, God's will is sovereign, and He not only controls the orbiting planets but manipulates the will of each and every individual as He sees fit. As Prime Mover, all causality flows from Him, and all events in the universe are foreordained. Sanguinary conflicts raged over the precise meaning of predestination or free will, from the time of Augustine through the Carolingian quarrels. Wycliffe, Huss, Luther, Calvin—all contributed to the debate.

No end of interpreters attempted to reconcile determinism with freedom of will. One ingenious view held that God did indeed determine the affairs of the universe, but that with respect to the free will of the individual, He never demanded a specific action. He merely preset the range of options available to the human decision-maker. Free will downstairs operated only within the limits of a menu determined upstairs.

In the secular culture of the Machine Age, hard-line determinism has more or less held sway even after the challenges of Heisenberg and the "uncertaintists." Even today, thinkers such as René Thom reject the idea of chance as illusory and inherently unscientific.

Faced with such philosophical stonewalling, some defenders of free will, spontaneity, and ultimate uncertainty, especially the existentialists, have taken equally uncompromising stands. (For Sartre, the human being was "completely and always free," though even Sartre, in certain writings, recognized practical limitations on this freedom.)

Two things seem to be happening to contemporary concepts of chance and determinism. To begin with, they are becoming more complex. As Edgar Morin, a leading French sociologist-turned-epistemologist, has written:

"Let us not forget that the problem of determinism has changed over the course of a century. . . . In place of the idea of sovereign, anonymous, permanent laws directing all things in nature there has been substituted the idea of laws of interaction. . . . There is more: the problem of determinism has become that of the order of the universe. Order means that there are other things besides 'laws': that there are constraints, invariances, constancies, regularities in our universe. . . . In place of the homogenizing and anonymous view of the old determinism, there has been substituted a diversifying and evolutive view of *determinations*."

And as the concept of determinism has grown richer, new efforts have been made to recognize the co-presence of both chance and necessity, not with one subordinate to the other, but as full partners in a universe that is simultaneously organizing and de-organizing itself.

It is here that Prigogine and Stengers enter the arena. For they have taken the argument a step farther. They not only demonstrate (persuasively to me, though not to critics like the mathematician, René Thom) that both determinism and chance operate, they also attempt to show how the two fit together.

Thus, according to the theory of change implied in the idea of dissipative structures, when fluctuations force an existing system into a far-from-equilibrium condition and threaten its structure, it approaches a critical moment or bifurcation point. At this point, according to the authors, it is inherently impossible to determine in advance the next state of the system. Chance nudges what remains of the system down a new path of development. And once that path is chosen (from among many), determinism takes over again until the next bifurcation point is reached.

Here, in short, we see chance and necessity not as irreconcilable opposites, but each playing its role as a partner in destiny.

Yet another synthesis is achieved.

When we bring reversible time and irreversible time, disorder and order, physics and biology, chance and necessity all into the same novel frame, and stipulate their interrelationships, we have made a grand statement—arguable, no doubt, but in this case both powerful and majestic.

Yet this accounts only in part for the excitement occasioned by *Order Out of Chaos*. For this sweeping synthesis, as I have suggested, has strong social and even political overtones. Just as the Newtonian model gave rise to analogies in politics, diplomacy, and other spheres seemingly remote from science, so, too, does the Prigoginian model lend itself to analogical extension.

By offering rigorous ways of modeling qualitative change, for example, they shed light on the concept of revolution. By explaining how successive instabilities give rise to transformatory change, they illuminate organization theory. They throw a fresh light, as well, on certain psychological processes—innovation, for example, which the authors see as associated with "nonaverage" behavior of the kind that arises under nonequilibrium conditions.

Even more significant, perhaps, are the implications for the study of collective behavior. Prigogine and Stengers caution against leaping to genetic or sociobiological explanations for puzzling social behavior. Many things that are attributed to biological pre-wiring are not produced by selfish, determinist genes, but rather by social interactions under nonequilibrium conditions.

(In one recent study, for instance, ants were divided into two categories: One consisted of hard workers, the other of inactive or "lazy" ants. One might overhastily trace such traits to genetic predisposition. Yet the study found that if the system were shattered by separating the two groups from one another, each in turn developed its own subgroups of hard workers and idlers. A significant percentage of the "lazy" ants suddenly turned into hardworking Stakhanovites.)

Not surprisingly, therefore, the ideas behind this remarkable book are beginning to be researched in economics, urban studies, human geography, ecology, and many other disciplines.

No one—not even its authors—can appreciate the full implications of a work as crowded with ideas as *Order Out of Chaos*. Each reader will no doubt come away puzzled by some passages (a few are simply too technical for the reader without scientific training); startled or stimulated by others (as their implications strike home); occasionally skeptical; yet intellectually enriched by the whole. And if one measure of a book is the degree to which it generates good questions, this one is surely successful.

Here are just a couple that have haunted me.

How, outside a laboratory, might one define a "fluctuation"? What, in Prigoginian terms, does one mean by "cause" or "effect"? And when the authors speak of molecules communicating with one another to achieve coherent, synchronized change, one may assume they are not anthropomorphizing. But they raise for me a host of intriguing issues about whether all parts of the environment are signaling all the time, or only intermittently; about the indirect, second, and *n*th order communication that takes place, permitting a molecule or an organism to respond to signals which it cannot sense for lack of the necessary receptors. (A signal sent by the environment that is undetectable by A may be received by B and converted into a different kind of signal that A *is* properly equipped to receive—so that B serves as a relay/converter, and A responds to an environmental change that has been signaled to it via second-order communication.)

In connection with time, what do the authors make of the idea put forward by Harvard astronomer David Layzer, that we might conceive of three distinct "arrows of time"—one based on the continued expansion of the universe since the Big Bang; one based on entropy; and one based on biological and historical evolution?

Another question: How revolutionary was the Newtonian revolution? Taking issue with some historians, Prigogine and Stengers point out the continuity of Newton's ideas with alchemy and religious notions of even earlier vintage. Some readers might conclude from this that the rise of Newtonianism was neither abrupt nor revolutionary. Yet, to my mind, the Newtonian breakthrough should not be seen as a linear outgrowth of these earlier ideas. Indeed, it seems to me that the theory of change developed in *Order Out of Chaos* argues against just such a "continuist" view.

Even if Newtonianism was derivative, this doesn't mean that the internal structure of the Newtonian world-model was actually the same or that it stood in the same relationship to its external environment.

The Newtonian system arose at a time when feudalism in Western Europe was crumbling—when the social system was, so to speak, far from equilibrium. The model of the universe proposed by the classical scientists (even if partially derivative) was applied analogously to new fields and disseminated successfully, not just because of its scientific power or "rightness," but also because an emergent industrial society based on revolutionary principles provided a particularly receptive environment for it.

As suggested earlier, machine civilization, in searching for an explanation of itself in the cosmic order of things, seized upon the Newtonian model and rewarded those who further developed it. It is not only in chemical beakers that we find auto-catalysis, as the authors would be the first to contend. For these reasons, it still makes sense to me to regard the Newtonian knowledge system as, itself, a "cultural dissipative structure" born of social fluctuation.

Ironically, as I've said, I believe their own ideas are central to the latest revolution in science, and I cannot help but see these ideas in relationship to the demise of the Machine Age and the rise of what I have called a "Third Wave" civilization. Applying their own terminology, we might characterize today's breakdown of industrial or "Second Wave" society as a civilizational "bifurcation," and the rise of a more differentiated, "Third Wave" society as a leap to a new "dissipative structure" on a world scale. And if we accept this analogy, might we not look upon the leap from Newtonianism to Prigoginianism in the same way? Mere analogy, no doubt. But illuminating, nevertheless.

Finally, we come once more to the ever-challenging issue of chance and necessity. For if Prigogine and Stengers are right and chance plays its role at or near the point of bifurcation, after which deterministic processes take over once more until the next bifurcation, are they not embedding chance, itself, within a deterministic framework? By assigning a particular role to chance, don't they de-chance it?

This question, however, I had the pleasure of discussing with Prigogine, who smiled over dinner and replied, "Yes. That would be true. But, of course, we can never determine when the next bifurcation will arise." Chance rises phoenix-like once more.

Order out of Chaos is a brilliant, demanding, dazzling book—challenging for all and richly rewarding for the attentive reader. It is a book to study, to savor, to reread—and to question yet again. It places science and humanity back in a world in which *ceteris paribus* is a myth—a world in which other things are seldom held steady, equal, or unchanging. In short, it projects science into today's revolutionary world of instability, disequilibrium, and turbulence. In so doing, it serves the highest creative function—it helps *us* create fresh order.

PREFACE

MAN'S NEW DIALOGUE WITH NATURE

Our vision of nature is undergoing a radical change toward the multiple, the temporal, and the complex. For a long time a mechanistic world view dominated Western science. In this view the world appeared as a vast automaton. We now understand that we live in a pluralistic world. It is true that there are phenomena that appear to us as deterministic and reversible, such as the motion of a frictionless pendulum or the motion of the earth around the sun. Reversible processes do not know any privileged direction of time. But there are also irreversible processes that involve an arrow of time. If you bring together two liquids such as water and alcohol, they tend to mix in the forward direction of time as we experience it. We never observe the reverse process, the spontaneous separation of the mixture into pure water and pure alcohol. This is therefore an irreversible process. All of chemistry involves such irreversible processes.

Obviously, in addition to deterministic processes, there must be an element of probability involved in some basic processes, such as, for example, biological evolution or the evolution of human cultures. Even the scientist who is convinced of the validity of deterministic descriptions would probably hesitate to imply that at the very moment of the Big Bang, the moment of the creation of the universe as we know it, the date of the publication of this book was already inscribed in the laws of nature. In the classical view the basic processes of nature were considered to be deterministic and reversible. Processes involving randomness or irreversibility were considered only exceptions. Today we see everywhere the role of irreversible processes, of fluctuations.

Although Western science has stimulated an extremely fruitful dialogue between man and nature, some of its cultural consequences have been disastrous. The dichotomy between the "two cultures" is to a large extent due to the conflict between the atemporal view of classical science and the time-oriented view that prevails in a large part of the social sciences and humanities. But in the past few decades, something very dramatic has been happening in science, something as unexpected as the birth of geometry or the grand vision of the cosmos as expressed in Newton's work. We are becoming more and more conscious of the fact that on all levels, from elementary particles to cosmology, randomness and irreversibility play an ever-increasing role. *Science is rediscovering time*. It is this conceptual revolution that this book sets out to describe.

This revolution is proceeding on all levels, on the level of elementary particles, in cosmology, and on the level of so-called macroscopic physics, which comprises the physics and chemistry of atoms and molecules either taken individually or considered globally as, for example, in the study of liquids or gases. It is perhaps particularly on this macroscopic level that the reconceptualization of science is most easy to follow. Classical dynamics and modern chemistry are going through a period of drastic change. If one asked a physicist a few years ago what physics permits us to explain and which problems remain open, he would have answered that we obviously do not have an adequate understanding of elementary particles or of cosmological evolution but that our knowledge of things in between was pretty satisfactory. Today a growing minority, to which we belong, would not share this optimism: we have only begun to understand the level of nature on which we live, and this is the level on which we have concentrated in this book.

To appreciate the reconceptualization of physics taking place today, we must put it in proper historical perspective. The history of science is far from being a linear unfolding that corresponds to a series of successive approximations toward some intrinsic truth. It is full of contradictions, of unexpected turning points. We have devoted a large portion of this book to the historical pattern followed by Western science, starting with Newton three centuries ago. We have tried to place the history of science in the frame of the history of ideas to integrate it in the evolution of Western culture during the past three centuries. Only in this way can we appreciate the unique moment in which we are presently living.

Our scientific heritage includes two basic questions to which till now no answer was provided. One is the relation between disorder and order. The famous law of increase of entropy describes the world as evolving from order to disorder; still, biological or social evolution shows us the complex emerging from the simple. How is this possible? How can structure arise from disorder? Great progress has been realized in this question. We know now that nonequilibrium, the flow of matter and energy, may be a source of order.

But there is the second question, even more basic: classical or quantum physics describes the world as reversible, as static. In this description there is no evolution, neither to order nor to disorder; the "information," as may be defined from dynamics, remains constant in time. Therefore there is an obvious contradiction between the static view of dynamics and the evolutionary paradigm of thermodynamics. What is irreversibility? What is entropy? Few questions have been discussed more often in the course of the history of science. We begin to be able to give some answers. Order and disorder are complicated notions: the units involved in the static description of dynamics are not the same as those that have to be introduced to achieve the evolutionary paradigm as expressed by the growth of entropy. This transition leads to a new concept of matter, matter that is "active," as matter leads to irreversible processes and as irreversible processes organize matter.

The evolutionary paradigm, including the concept of entropy, has exerted a considerable fascination that goes far beyond science proper. We hope that our unification of dynamics and thermodynamics will bring out clearly the radical novelty of the entropy concept in respect to the mechanistic world view. Time and reality are closely related. For humans, reality is embedded in the flow of time. As we shall see, the irreversibility of time is itself closely connected to entropy. To make time flow backward we would have to overcome an infinite entropy barrier.

Traditionally science has dealt with universals, humanities with particulars. The convergence of science and humanities was emphasized in the French title of our book, *La Nouvelle*

Alliance, published by Gallimard, Paris, in 1979. However, we have not succeeded in finding a proper English equivalent of this title. Furthermore, the text we present here differs from the French edition, especially in Chapters VII through IX. Although the origin of structures as the result of nonequilibrium processes was already adequately treated in the French edition (as well as in the translations that followed), we had to entirely rewrite the third part, which deals with our new results concerning the roots of time as well as with the formulation of the evolutionary paradigm in the frame of the physical sciences.

This is all quite recent. The reconceptualization of physics is far from being achieved. We have decided, however, to present the situation as it seems to us today. We have a feeling of great intellectual excitement: we begin to have a glimpse of the road that leads from being to becoming. As one of us has devoted most of his scientific life to this problem, he may perhaps be excused for expressing his feeling of satisfaction, of aesthetic achievement, which he hopes the reader will share. For too long there appeared a conflict between what seemed to be eternal, to be out of time, and what was in time. We see now that there is a more subtle form of reality involving both time and eternity.

This book is the outcome of a collective effort in which many colleagues and friends have been involved. We cannot thank them all individually. We would like, however, to single out what we owe to Erich Jantsch, Aharon Katchalsky, Pierre Résibois, and Léon Rosenfeld, who unfortunately are no longer with us. We have chosen to dedicate this book to their memory.

We want also to acknowledge the continuous support we have received from the Instituts Internationaux de Physique et de Chimie, founded by E. Solvay, and from the Robert A. Welch Foundation.

The human race is in a period of transition. Science is likely to play an important role at this moment of demographic explosion. It is therefore more important than ever to keep open the channels of communication between science and society. The present development of Western science has taken it outside the cultural environment of the seventeenth century, in which it was born. We believe that science today carries a universal message that is more acceptable to different cultural traditions.

During the past decades Alvin Toffler's books have been important in bringing to the attention of the public some features of the "Third Wave" that characterizes our time. We are therefore grateful to him for having written the Foreword to the English-language version of our book. English is not our native language. We believe that to some extent every language provides a different way of describing the common reality in which we are embedded. Some of these characteristics will survive even the most careful translation. In any case, we are most grateful to Joseph Early, Ian MacGilvray, Carol Thurston, and especially to Carl Rubino for their help in the preparation of this English-language version. We would also like to express our deep thanks to Pamela Pape for the careful typing of the successive versions of the manuscript.

INTRODUCTION
THE CHALLENGE TO SCIENCE

1

It is hardly an exaggeration to state that one of the greatest dates in the history of mankind was April 28, 1686, when Newton presented his *Principia* to the Royal Society of London. It contained the basic laws of motion together with a clear formulation of some of the fundamental concepts we still use today, such as mass, acceleration, and inertia. The greatest impact was probably made by Book III of the *Principia,* titled *The System of the World,* which included the universal law of gravitation. Newton's contemporaries immediately grasped the unique importance of his work. Gravitation became a topic of conversation both in London and Paris.

Three centuries have now elapsed since Newton's *Principia*. Science has grown at an incredible speed, permeating the life of all of us. Our scientific horizon has expanded to truly fantastic proportions. On the microscopic scale, elementary particle physics studies processes involving physical dimensions of the order of 10^{-15} cm and times of the order of 10^{-22} second. On the other hand, cosmology leads us to times of the order of 10^{10} years, the "age of the universe." Science and technology are closer than ever. Among other factors, new biotechnologies and the progress in information techniques promise to change our lives in a radical way.

Running parallel to this quantitative growth are deep qualitative changes whose repercussions reach far beyond science proper and affect the very image of nature. The great founders of Western science stressed the universality and the eternal character of natural laws. They set out to formulate general schemes that would coincide with the very ideal of rationality.

As Roger Hausheer says in his fine introduction to Isaiah Berlin's *Against the Current,* "They sought all-embracing schemas, universal unifying frameworks, within which everything that exists could be shown to be systematically—i.e., logically or causally—interconnected, vast structures in which there should be no gaps left open for spontaneous, unattended developments, where everything that occurs should be, at least in principle, wholly explicable in terms of immutable general laws."[1]

The story of this quest is indeed a dramatic one. There were moments when this ambitious program seemed near completion. A fundamental level from which all other properties of matter could be deduced seemed to be in sight. Such moments can be associated with the formulation of Bohr's celebrated atomic model, which reduced matter to simple planetary systems formed by electrons and protons. Another moment of great suspense came when Einstein hoped to condense all the laws of physics into a single "unified field theory." Great progress has indeed been realized in the unification of some of the basic forces found in nature. Still, the fundamental level remains elusive. Wherever we look we find evolution, diversification, and instabilities. Curiously, this is true on all levels, in the field of elementary particles, in biology, and in astrophysics, with the expanding universe and the formation of black holes.

As we said in the Preface, our vision of nature is undergoing a radical change toward the multiple, the temporal, and the complex. Curiously, the unexpected complexity that has been discovered in nature has not led to a slowdown in the progress of science, but on the contrary to the emergence of new conceptual structures that now appear as essential to our understanding of the physical world—the world that includes us. It is this new situation, which has no precedent in the history of science, that we wish to analyze in this book.

The story of the transformation of our conceptions about science and nature can hardly be separated from another story, that of the feelings aroused by science. With every new intellectual program always come new hopes, fears, and expectations. In classical science the emphasis was on time-independent laws. As we shall see, once the particular state of a system has been measured, the reversible laws of classical science are supposed to determine its future, just as they had determined its past. It is natural that this quest for an eternal truth behind changing phenomena aroused enthusiasm. But it also came as a shock that nature described in this way was in fact debased: by the very success of science, nature was shown to be an automaton, a robot.

The urge to reduce the diversity of nature to a web of illusions has been present in Western thought since the time of Greek atomists. Lucretius, following his masters Democritus and Epicurus, writes that the world is "just" atoms and void and urges us to look for the hidden behind the obvious: "Still, lest you happen to mistrust my words, because the eye cannot perceive prime bodies, hear now of particles you must admit exist in the world and yet cannot be seen."[2]

Yet it is well known that the driving force behind the work of the Greek atomists was not to debase nature but to free men from fear, the fear of any supernatural being, of any order that would transcend that of men and nature. Again and again Lucretius repeats that we have nothing to fear, that the essence of the world is the ever-changing associations of atoms in the void.

Modern science transmuted this fundamentally ethical stance into what seemed to be an established truth; and this truth, the reduction of nature to atoms and void, in turn gave rise to what Lenoble[3] has called the "anxiety of modern men." How can we recognize ourselves in the random world of the atoms? Must science be defined in terms of rupture between man and nature? "All bodies, the firmament, the stars, the earth and its kingdoms are not equal to the lowest mind; for mind knows all this in itself and these bodies nothing."[4] This "Pensée" by Pascal expresses the same feeling of alienation we find among contemporary scientists such as Jacques Monod:

> Man must at last finally awake from his millenary dream; and in doing so, awake to his total solitude, his fundamental isolation. Now does he at last realize that, like a gypsy, he lives on the boundary of an alien world. A world that is deaf to his music, just as indifferent to his hopes as it is to his suffering or his crimes.[5]

This is a paradox. A brilliant breakthrough in molecular biology, the deciphering of the genetic code, in which Monod actively participated, ends upon a tragic note. This very progress, we are told, makes us the gypsies of the universe. How can we explain this situation? Is not science a way of communication, a dialogue with nature?

In the past, strong distinctions were frequently made between man's world and the supposedly alien natural world. A famous passage by Vico in *The New Science* describes this most vividly:

> . . . in the night of thick darkness enveloping the earliest antiquity, so remote from ourselves, there shines the eternal and never failing light of a truth beyond all question: that the world of civil society has certainly been made by men, and that its principles are therefore to be found within the modifications of our own human mind. Whoever reflects on this cannot but marvel that the philosophers should have bent all their energies to the study of the world of nature, which, since God made it, He alone knows; and that they should have neglected the study of the world of nations, or civil world, which, since men had made it, men could come to know.[6]

Present-day research leads us farther and farther away from the opposition between man and the natural world. It will be one of the main purposes of this book to show, instead of rupture and opposition, the growing coherence of our knowledge of man and nature.

2

In the past, the questioning of nature has taken the most diverse forms. Sumer discovered writing; the Sumerian priests speculated that the future might be written in some hidden way in the events taking place around us in the present. They even systematized this belief, mixing magical and rational elements.[7] In this sense we may say that Western science, which originated in the seventeenth century, only opened a new chapter in the everlasting dialogue between man and nature.

Alexandre Koyré[8] has defined the innovation brought about by modern science in terms of "experimentation." Modern science is based on the discovery of a new and specific form of communication with nature—that is, on the conviction that nature responds to experimental interrogation. How can we define more precisely the experimental dialogue? Experimentation does not mean merely the faithful observation of facts as they occur, nor the mere search for empirical connections between phenomena, but presupposes a systematic interaction between theoretical concepts and observation.

In hundreds of different ways scientists have expressed their amazement when, on determining the right question, they discover that they can see how the puzzle fits together. In this sense, science is like a two-partner game in which we have to guess the behavior of a reality unrelated to our beliefs, our ambitions, or our hopes. Nature cannot be forced to say anything we want it to. Scientific investigation is not a monologue. It is precisely the risk involved that makes this game exciting.

But the uniqueness of Western science is far from being exhausted by such methodological considerations. When Karl Popper discussed the normative description of scientific rationality, he was forced to admit that in the final analysis rational science owes its existence to its success; the scientific method is applicable only by virtue of the astonishing points of agreement between preconceived models and experimental results.[9] Science is a risky game, but it seems to have discovered questions to which nature provides consistent answers.

The success of Western science is an historical fact, unpredictable a priori, but which cannot be ignored. The surprising success of modern science has led to an irreversible transformation of our relations with nature. In this sense, the term "scientific revolution" can legitimately be used. The history of mankind has been marked by other turning points, by other singular conjunctions of circumstances leading to irreversible changes. One such crucial event is known as the "Neolithic revolution." But there, just as in the case of the "choices" marking biological evolution, we can at present only proceed by guesswork, while there is a wealth of information concerning decisive episodes in the evolution of science. The so-called

"Neolithic revolution" took thousands of years. Simplifying somewhat, we may say the scientific revolution started only three centuries ago. We have what is perhaps a unique opportunity to apprehend the specific and intelligible mixture of "chance" and "necessity" marking this revolution.

Science initiated a successful dialogue with nature. On the other hand, the first outcome of this dialogue was the discovery of a silent world. This is the paradox of classical science. It revealed to men a dead, passive nature, a nature that behaves as an automaton which, once programmed, continues to follow the rules inscribed in the program. In this sense the dialogue with nature isolated man from nature instead of bringing him closer to it. A triumph of human reason turned into a sad truth. It seemed that science debased everything it touched.

Modern science horrified both its opponents, for whom it appeared as a deadly danger, and some of its supporters, who saw in man's solitude as "discovered" by science the price we had to pay for this new rationality.

The cultural tension associated with classical science can be held at least partly responsible for the unstable position of science within society; it led to an heroic assumption of the harsh implications of rationality, but it led also to violent rejection. We shall return later to present-day antiscience movements. Let us take an earlier example—the irrationalist movement in Germany in the 1920s that formed the cultural background to quantum mechanics.[10] In opposition to science, which was identified with a set of concepts such as causality, determinism, reductionism, and rationality, there was a violent upsurge of ideas denied by science but seen as the embodiment of the fundamental irrationality of nature. Life, destiny, freedom, and spontaneity thus became manifestations of a shadowy underworld impenetrable to reason. Without going into the peculiar sociopolitical context to which it owed its vehement nature, we can state that this rejection illustrates the risks associated with classical science. By admitting only a subjective meaning for a set of experiences men believe to be significant, science runs the risk of transferring these into the realm of the irrational, bestowing upon them a formidable power.

As Joseph Needham has emphasized, Western thought has

always oscillated between the world as an automaton and a theology in which God governs the universe. This is what Needham calls the "characteristic European schizophrenia."[11] In fact, these visions are connected. An automaton needs an external god.

Do we really have to make this tragic choice? Must we choose between a science that leads to alienation and an antiscientific metaphysical view of nature? We think such a choice is no longer necessary, since the changes that science is undergoing today lead to a radically new situation. This recent evolution of science gives us a unique opportunity to reconsider its position in culture in general. Modern science originated in the specific context of the European seventeenth century. We are now approaching the end of the twentieth century, and it seems that some more *universal message* is carried by science, a message that concerns the interaction of man and nature as well as of man with man.

3

What are the assumptions of classical science from which we believe science has freed itself today? Generally those centering around the basic conviction that at some level *the world is simple* and is governed by time-reversible fundamental laws. Today this appears as an excessive simplification. We may compare it to reducing buildings to piles of bricks. Yet out of the same bricks we may construct a factory, a palace, or a cathedral. It is on the level of the building as a whole that we apprehend it as a creature of time, as a product of a culture, a society, a style. But there is the additional and obvious problem that, since there is no one to build nature, we must give to its very "bricks"—that is, to its microscopic activity—a description that accounts for this building process.

The quest of classical science is in itself an illustration of a dichotomy that runs throughout the history of Western thought. Only the immutable world of ideas was traditionally recognized as "illuminated by the sun of the intelligible," to use Plato's expression. In the same sense, only eternal laws were seen to express scientific rationality. Temporality was looked down upon as an illusion. This is no longer true today.

We have discovered that far from being an illusion, irreversibility plays an essential role in nature and lies at the origin of most processes of self-organization. We find ourselves in a world in which reversibility and determinism apply only to limiting, simple cases, while irreversibility and randomness are the rules.

The denial of time and complexity was central to the cultural issues raised by the scientific enterprise in its classical definition. The challenge of these concepts was also decisive for the metamorphosis of science we wish to describe. In his great book *The Nature of the Physical World,* Arthur Eddington[12] introduced a distinction between primary and secondary laws. "Primary laws" control the behavior of single particles, while "secondary laws" are applicable to collections of atoms or molecules. To insist on secondary laws is to emphasize that the description of elementary behaviors is not sufficient for understanding a system as a whole. An outstanding case of a secondary law is, in Eddington's view, the second law of thermodynamics, the law that introduces the "arrow of time" in physics. Eddington writes: "From the point of view of philosophy of science the conception associated with entropy must, I think, be ranked as the great contribution of the nineteenth century to scientific thought. It marked a reaction from the view that everything to which science need pay attention is discovered by a microscopic dissection of objects."[13] This trend has been dramatically amplified today.

It is true that some of the greatest successes of modern science are discoveries at the microscopic level, that of molecules, atoms, or elementary particles. For example, molecular biology has been immensely successful in isolating specific molecules that play a central role in the mechanism of life. In fact, this success has been so overwhelming that for many scientists the aim of research is identified with this "microscopic dissection of objects," to use Eddington's expression. However, the second law of thermodynamics presented the first challenge to a concept of nature that would explain away the complex and reduce it to the simplicity of some hidden world. Today interest is shifting from substance to relation, to communication, to time.

This change of perspective is not the result of some arbitrary decision. In physics it was forced upon us by new discoveries no one could have foreseen. Who would have expected that most (and perhaps all) elementary particles would prove to be unstable? Who would have expected that with the experimental confirmation of an expanding universe we could conceive of the history of the world as a whole?

At the end of the twentieth century we have learned to understand better the meaning of the two great revolutions that gave shape to the physics of our time, quantum mechanics and relativity. They started as attempts to correct classical mechanics and to incorporate into it the newly found universal constants. Today the situation has changed. Quantum mechanics has given us the theoretical frame to describe the incessant transformations of particles into each other. Similarly, general relativity has become the basic theory in terms of which we can describe the thermal history of our universe in its early stages.

Our universe has a pluralistic, complex character. Structures may disappear, but also they may appear. Some processes are, as far as we know, well described by deterministic equations, but others involve probabilistic processes.

How then can we overcome the apparent contradiction between these concepts? We are living in a single universe. As we shall see, we are beginning to appreciate the meaning of these problems. Moreover, the importance we now give to the various phenomena we observe and describe is quite different from, even opposite to, what was suggested by classical physics. There the basic processes, as we mentioned, are considered as deterministic and reversible. Processes involving randomness or irreversibility are considered to be exceptions. Today we see everywhere the role of irreversible processes, of fluctuations. The models considered by classical physics seem to us to occur only in limiting situations such as we can create artificially by putting matter into a box and then waiting till it reaches equilibrium.

The artificial may be deterministic and reversible. The natural contains essential elements of randomness and irreversibility. This leads to a new view of matter in which matter is no longer the passive substance described in the mechanistic world view but is associated with spontaneous activity. This change is so profound that, as we stated in our Preface, we can really speak about a new dialogue of man with nature.

4

This book deals with the conceptual transformation of science from the Golden Age of classical science to the present. To describe this transformation we could have chosen many roads. We could have studied the problems of elementary particles. We could have followed recent fascinating developments in astrophysics. These are the subjects that seem to delimit the frontiers of science. However, as we stated in our Preface, over the past years so many new features of nature at our level have been discovered that we decided to concentrate on this intermediate level, on problems that belong mainly to our macroscopic world, which includes atoms, molecules, and especially biomolecules. Still it is important to emphasize that the evolution of science proceeds on somewhat parallel lines at every level, be it that of elementary particles, chemistry, biology, or cosmology. On every scale self-organization, complexity, and time play a new and unexpected role.

Therefore, our aim is to examine the significance of three centuries of scientific progress from a definite viewpoint. There is certainly a subjective element in the way we have chosen our material. The problem of time is really the center of the research that one of us has been pursuing all his life. When as a young student at the University of Brussels he came into contact with physics and chemistry for the first time, he was astonished that science had so little to say about time, especially since his earlier education had centered mainly around history and archaeology. This surprise could have led him to two attitudes, both of which we find exemplified in the past: one would have been to discard the problem, since classical science seemed to have no place for time; and the other would have been to look for some other way of apprehending nature, in which time would play a different, more basic role. This is the path Bergson and Whitehead, to mention only two philosophers of our century, chose. The first position would be a "positivistic" one, the second a "metaphysical" one.

There was, however, a third path, which was to ask whether the simplicity of the temporal evolution traditionally considered in physics and chemistry was due to the fact that attention was paid mainly to some very simplified situations, to heaps of bricks in contrast with the cathedral to which we have alluded.

This book is divided into three parts. The first part deals with the triumph of classical science and the cultural consequences of this triumph. Initially, science was greeted with enthusiasm. We shall then describe the cultural polarization that occurred as a result of the *existence* of classical science and its astonishing success. Is this success to be accepted as such, perhaps limiting its implications, or must the scientific method itself be rejected as partial or illusory? Both choices lead to the same result—the collision between what has often been called the "two cultures," science and the humanities.

These questions have played a basic role in Western thought since the formulation of classical science. Again and again we come to the problem, "How to choose?" Isaiah Berlin has rightly seen in this question the beginning of the schism between the sciences and the humanities:

> The specific and unique versus the repetitive and the universal, the concrete versus the abstract, perpetual movement versus rest, the inner versus the outer, quality versus quantity, culture-bound versus timeless principles, mental strife and self-transformation as a permanent condition of man versus the possibility (and desirability) of peace, order, final harmony and the satisfaction of all rational human wishes—these are some of the aspects of the contrast.[14]

We have devoted much space to classical mechanics. Indeed, in our view this is the best vantage point from which we may contemplate the present-day transformation of science. Classical dynamics seems to express in an especially clear and striking way the static view of nature. Here time apparently is reduced to a parameter, and future and past become equivalent. It is true that quantum theory has raised many new problems not covered by classical dynamics but it has nevertheless retained a number of the conceptual positions of classical dynamics, particularly as far as time and process are concerned.

As early as at the beginning of the nineteenth century, precisely when classical science was triumphant, when the Newtonian program dominated French science and the latter dominated Europe, the first threat to the Newtonian construction loomed into sight. In the second part of our study we shall follow the development of the science of heat, this rival to Newton's science of gravity, starting from the first gauntlet thrown down when Fourier formulated the law governing the propagation of heat. It was, in fact, the first quantitative description of something inconceivable in classical dynamics—an irreversible process.

The two descendants of the science of heat, the science of energy conversion and the science of heat engines, gave birth to the first "nonclassical" science—thermodynamics. The most original contribution of thermodynamics is the celebrated second law, which introduced into physics the arrow of time. This introduction was part of a more global intellectual move. The nineteenth century was really the century of evolution; biology, geology, and sociology emphasized processes of becoming, of increasing complexity. As for thermodynamics, it is based on the distinction of two types of processes: reversible processes, which are independent of the direction of time, and irreversible processes, which depend on the direction of time. We shall see examples later. It was in order to distinguish the two types of processes that the concept of entropy was introduced, since entropy increases only because of the irreversible processes.

During the nineteenth century the final state of thermodynamic evolution was at the center of scientific research. This was equilibrium thermodynamics. Irreversible processes were looked down on as nuisances, as disturbances, as subjects not worthy of study. Today this situation has completely changed. We now know that far from equilibrium, new types of structures may originate spontaneously. In far-from-equilibrium conditions we may have transformation from disorder, from thermal chaos, into order. New dynamic states of matter may originate, states that reflect the interaction of a given system with its surroundings. We have called these new structures *dissipative structures* to emphasize the constructive role of dissipative processes in their formation.

This book describes some of the methods that have been

developed in recent years to deal with the appearance and evolution of dissipative structures. Here we find the key words that run throughout this book like leitmotivs: nonlinearity, instability, fluctuations. They have begun to permeate our view of nature even beyond the fields of physics and chemistry proper.

We cited Isaiah Berlin when we discussed the opposition between the sciences and the humanities. He opposed the specific and unique to the repetitive and the universal. The remarkable feature is that when we move from equilibrium to far-from-equilibrium conditions, we move away from the repetitive and the universal to the specific and the unique. Indeed, the laws of equilibrium are universal. Matter near equilibrium behaves in a "repetitive" way. On the other hand, far from equilibrium there appears a variety of mechanisms corresponding to the possibility of occurrence of various types of dissipative structures. For example, far from equilibrium we may witness the appearance of chemical clocks, chemical reactions which behave in a coherent, rhythmical fashion. We may also have processes of self-organization leading to nonhomogeneous structures to nonequilibrium crystals.

We would like to emphasize the unexpected character of this behavior. Every one of us has an intuitive view of how a chemical reaction takes place; we imagine molecules floating through space, colliding, and reappearing in new forms. We see chaotic behavior similar to what the atomists described when they spoke about dust dancing in the air. But in a chemical clock the behavior is quite different. Oversimplifying somewhat, we can say that in a chemical clock all molecules change their chemical identity *simultaneously,* at regular time intervals. If the molecules can be imagined as blue or red, we would see their change of color following the rhythm of the chemical clock reaction.

Obviously such a situation can no longer be described in terms of chaotic behavior. A new type of order has appeared. We can speak of a new coherence, of a mechanism of "communication" among molecules. But this type of communication can arise only in far-from-equilibrium conditions. It is quite interesting that such communication seems to be the rule in the world of biology. It may in fact be taken as the very basis of the definition of a biological system.

In addition, the type of dissipative structure depends critically on the conditions in which the structure is formed. External fields such as the gravitational field of earth, as well as the magnetic field, may play an essential role in the selection mechanism of self-organization.

We begin to see how, starting from chemistry, we may build complex structures, complex forms, some of which may have been the precursors of life. What seems certain is that these far-from-equilibrium phenomena illustrate an essential and unexpected property of matter: physics may henceforth describe structures as adapted to outside conditions. We meet in rather simple chemical systems a kind of prebiological adaptation mechanism. To use somewhat anthropomorphic language: in equilibrium matter is "blind," but in far-from-equilibrium conditions it begins to be able to perceive, to "take into account," in its way of functioning, differences in the external world (such as weak gravitational or electrical fields).

Of course, the problem of the origin of life remains a difficult one, and we do not think a simple solution is imminent. Still, from this perspective life no longer appears to oppose the "normal" laws of physics, struggling against them to avoid its normal fate—its destruction. On the contrary, life seems to express in a specific way the very conditions in which our biosphere is embedded, incorporating the nonlinearities of chemical reactions and the far-from-equilibrium conditions imposed on the biosphere by solar radiation.

We have discussed the concepts that allow us to describe the formation of dissipative structures, such as the theory of bifurcations. It is remarkable that near-bifurcations systems present large fluctuations. Such systems seem to "hesitate" among various possible directions of evolution, and the famous law of large numbers in its usual sense breaks down. A small fluctuation may start an entirely new evolution that will drastically change the whole behavior of the macroscopic system. The analogy with social phenomena, even with history, is inescapable. Far from opposing "chance" and "necessity," we now see both aspects as essential in the description of nonlinear systems far from equilibrium.

The first two parts of this book thus deal with two conflicting views of the physical universe: the static view of classical dynamics, and the evolutionary view associated with entropy.

A confrontation between these views has become unavoidable. For a long time this confrontation was postponed by considering irreversibility as an illusion, as an approximation; it was man who introduced time into a timeless universe. However, this solution in which irreversibility is reduced to an illusion or to approximations can no longer be accepted, since we know that irreversibility may be a source of order, of coherence, of organization.

We can no longer avoid this confrontation. It is the subject of the third part of this book. We describe traditional attempts to approach the problem of irreversibility first in classical and then in quantum mechanics. Pioneering work was done here, especially by Boltzmann and Gibbs. However, we can state that the problem was left largely unsolved. As Karl Popper relates it, it is a dramatic story: first, Boltzmann thought he had given an objective formulation to the new concept of time implied in the second law. But as a result of his controversy with Zermelo and others, he had to retreat.

> In the light of history—or in the darkness of history—Boltzmann was defeated, according to all accepted standards, though everybody accepts his eminence as a physicist. For he never succeeded in clearing up the status of his $\mathcal{H}$-theorem; nor did he explain entropy increase. . . . Such was the pressure that he lost faith in himself. . . .[15]

The problem of irreversibility still remains a subject of lively controversy. How is this possible one hundred fifty years after the discovery of the second law of thermodynamics? There are many aspects to this question, some cultural and some technical. There is a cultural component in the mistrust of time. We shall on several occasions cite the opinion of Einstein. His judgment sounds final: time (as irreversibility) is an illusion. In fact, Einstein was reiterating what Giordano Bruno had written in the sixteenth century and what had become for centuries the credo of science: "The universe is, therefore, one, infinite, immobile. . . . It does not move itself locally. . . . It does not generate itself. . . . It is not corruptible. . . . It is not alterable. . . ."[16] For a long time Bruno's vision dominated the scientific view of the Western world. It is therefore not surprising that the intrusion of irreversibility, coming mainly from the engineering sciences and physical chemistry, was received with mistrust. But there are technical reasons in addition to cultural ones. Every attempt to "derive" irreversibility from dynamics necessarily had to fail, because irreversibility is not a universal phenomenon. We can imagine situations that are strictly reversible, such as a pendulum in the absence of friction, or planetary motion. This failure has led to discouragement and to the feeling that, in the end, the whole concept of irreversibility has a subjective origin. We shall discuss all these problems at some length. Let us say here that today we can see this problem from a different point of view, since we now know that there are different classes of dynamic systems. The world is far from being homogeneous. Therefore the question can be put in different terms: What is the specific structure of dynamic systems that permits them to "distinguish" past and future? What is the minimum complexity involved?

Progress has been realized along these lines. We can now be more precise about the roots of time in nature. This has far-reaching consequences. The second law of thermodynamics, the law of entropy, introduced irreversibility into the macroscopic world. We now can understand its meaning on the microscopic level as well. As we shall see, the second law corresponds to a selection rule, to a restriction on initial conditions that is then propagated by the laws of dynamics. Therefore the second law introduces a new irreducible element into our description of nature. While it is consistent with dynamics, it cannot be derived from dynamics.

Boltzmann already understood that probability and irreversibility had to be closely related. Only when a system behaves in a sufficiently random way may the difference between past and future, and therefore irreversibility, enter into its description. Our analysis confirms this point of view. Indeed, what is the meaning of an arrow of time in a deterministic description of nature? If the future is already in some way contained in the present, which also contains the past, what is the meaning of an arrow of time? The arrow of time is a manifestation of the fact that the future is not given, that, as the French poet Paul Valêry emphasized, "time is construction."[17]

The experience of our everyday life manifests a radical difference between time and space. We can move from one point of space to another. However, we cannot turn time around. We cannot exchange past and future. As we shall see, this feeling of impossibility is now acquiring a precise scientific meaning. Permitted states are separated from states that are prohibited by the second law of thermodynamics by means of an infinite entropy barrier. There are other barriers in physics. One is the velocity of light, which in our present view limits the speed at which signals may be transmitted. It is essential that this barrier exist; if not, causality would fall to pieces. Similarly, the entropy barrier is the prerequisite for giving a meaning to communication. Imagine what would happen if our future would become the past for other people! We shall return to this later.

The recent evolution of physics has emphasized the reality of time. In the process new aspects of time have been uncovered. A preoccupation with time runs all through our century. Think of Einstein, Proust, Freud, Teilhard, Peirce, or Whitehead.

One of the most surprising results of Einstein's special theory of relativity, published in 1905, was the introduction of a local time associated with each observer. However, this local time remained a reversible time. Einstein's problem both in the special and the general theories of relativity was mainly that of the "communication" between observers, the way they could compare time intervals. But we can now investigate time in other conceptual contexts.

In classical mechanics time was a number characterizing the position of a point on its trajectory. But time may have a different meaning on a global level. When we look at a child and guess his or her age, this age is not located in any special part of the child's body. It is a global judgment. It has often been stated that science spatializes time. But we now discover that another point of view is possible. Consider a landscape and its evolution: villages grow, bridges and roads connect different regions and transform them. Space thus acquires a temporal dimension; following the words of geographer B. Berry, we have been led to study the "timing of space."

But perhaps the most important progress is that we now may see the problem of structure, of order, from a different perspective. As we shall show in Chapter VIII, from the point of view of dynamics, be it classical or quantum, there can be no one time-directed evolution. The "information" as it can be defined in terms of dynamics remains constant in time. This sounds paradoxical. When we mix two liquids, there would occur no "evolution" in spite of the fact that we cannot, without using some external device, undo the effect of the mixing. On the contrary, the entropy law describes the mixing as the evolution toward a "disorder," toward the most probable state. We can show now that there is no contradiction between the two descriptions, but to speak about information, or order, we have to redefine the units we are considering. The important new fact is that we now may establish precise rules to go from one type of unit to the other. In other words, we have achieved a microscopic formulation of the evolutionary paradigm expressed by the second law. As the evolutionary paradigm encompasses all of chemistry as well as essential parts of biology and the social sciences, this seems to us an important conclusion. This insight is quite recent. The process of reconceptualization occurring in physics is far from being complete. However, our intention is not to shed light on the definitive acquisitions of science, on its stable and well-established results. What we wish to do is emphasize the conceptual creativeness of scientific activity and the future prospects and new problems it raises. In any case, we now know that we are only at the beginning of this exploration. We shall not see the end of uncertainty or risk. Thus we have chosen to present things as we perceive them now, fully aware of how incomplete our answers are.

5

Erwin Schrödinger once wrote, to the indignation of many philosophers of science:

> . . . there is a tendency to forget that all science is bound up with human culture in general, and that scientific findings, even those which at the moment appear the most advanced and esoteric and difficult to grasp, are meaningless outside their cultural context. A theoretical science unaware that those of its constructs considered relevant and momentous are destined eventually to be framed in concepts and words that have a grip on the educated com-

munity and become part and parcel of the general world picture—a theoretical science, I say, where this is forgotten, and where the initiated continue musing to each other in terms that are, at best, understood by a small group of close fellow travellers, will necessarily be cut off from the rest of cultural mankind; in the long run it is bound to atrophy and ossify however virulently esoteric chat may continue within its joyfully isolated groups of experts.[18]

One of the main themes of this book is that of a strong interaction of the issues proper to culture as a whole and the internal conceptual problems of science in particular. We find questions about time at the very heart of science. Becoming, irreversibility—these are questions to which generations of philosophers have also devoted their lives. Today, when history—be it economic, demographic, or political—is moving at an unprecedented pace, new questions and new interests require us to enter into new dialogues, to look for a new coherence.

However, we know the progress of science has often been described in terms of rupture, as a shift away from concrete experience toward a level of abstraction that is increasingly difficult to grasp. We believe that this kind of interpretation is only a reflection, at the epistemological level, of the historical situation in which classical science found itself, a consequence of its inability to include in its theoretical frame vast areas of the relationship between man and his environment.

There doubtless exists an abstract development of scientific theories. However, the conceptual innovations that have been decisive for the development of science are not necessarily of this type. The rediscovery of time has roots both in the internal history of science and in the social context in which science finds itself today. Discoveries such as those of unstable elementary particles or of the expanding universe clearly belong to the internal history of science, but the general interest in nonequilibrium situations, in evolving systems, may reflect our feeling that humanity as a whole is today in a transition period. Many results we shall report in Chapters V and VI, for example those on oscillating chemical reactions, could have been discovered many years ago, but the study of these nonequilibrium problems was repressed in the cultural and ideological context of those times.

We are aware that asserting this receptiveness to cultural content runs counter to the traditional conception of science. In this view science develops by freeing itself from outmoded forms of understanding nature; it purifies itself in a process that can be compared to an "ascesis" of reason. But this in turn leads to the conclusion that science should be practiced only by communities living apart, uninvolved in mundane matters. In this view, the ideal scientific community should be protected from the pressures, needs, and requirements of society. Scientific progress ought to be an essentially autonomous process that any "outside" influence, such as the scientists's participation in other cultural, social, or economic activities, would merely disturb or delay.

This ideal of abstraction, of the scientist's withdrawal, finds an ally in still another ideal, this one concerning the vocation of a "true" researcher, namely, his desire to escape from worldly vicissitudes. Einstein describes the type of scientist who would find favor with the "Angel of the Lord" should the latter be given the task of driving from the "Temple of Science" all those who are "unworthy"—it is not stated in what respects. They are generally

> . . . rather odd, uncommunicative, solitary fellows, who despite these common characteristics resemble one another really less than the host of the banished.
>
> What led them into the Temple? . . . one of the strongest motives that lead men to art and science is flight from everyday life with its painful harshness and wretched dreariness, and from the fetters of one's own shifting desires. A person with a finer sensibility is driven to escape from personal existence and to the world of objective observing *(Schauen)* and understanding. This motive can be compared with the longing that irresistibly pulls the town-dweller away from his noisy, cramped quarters and toward the silent, high mountains, where the eye ranges freely through the still, pure air and traces the calm contours that seem to be made for eternity.
>
> With this negative motive there goes a positive one. Man seeks to form for himself, in whatever manner is suitable

> for him, a simplified and lucid image of the world *(Bild der Welt),* and so to overcome the world of experience by striving to replace it to some extent by this image.[19]

The incompatibility between the ascetic beauty sought after by science, on the one hand, and the petty swirl of worldly experience so keenly felt by Einstein, on the other, is likely to be reinforced by another incompatibility, this one openly Manichean, between science and society, or, more precisely, between free human creativity and political power. In this case, it is not in an isolated community or in a temple that research would have to be carried out, but in a fortress, or else in a madhouse, as Duerrenmatt imagined in his play *The Physicists.*[20] There, three physicists discuss the ways and means of advancing physics while at the same time safeguarding mankind from the dire consequences that result when political powers appropriate the results of its progress. The conclusion they reach is that the only possible way is that which has already been chosen by one of them; they all decide to pretend to be mad, to hide in a lunatic asylum. At the end of the play, as Fate would have it, this last refuge is discovered to be an illusion. The director of the asylum, who has been spying on her patient, steals his results and seizes world power.

Duerrenmatt's play leads to a third conception of scientific activity: science progresses by reducing the complexity of reality to a hidden simplicity. What the physicist Moebius is trying to conceal in the madhouse is the fact that he has successfully solved the problem of gravitation, the unified theory of elementary particles, and, ultimately, the Principle of Universal Discovery, the source of absolute power. Of course, Duerrenmatt simplifies to make his point, yet it is commonly held that what is being sought in the "Temple of Science" is nothing less than the "formula" of the universe. The man of science, already portrayed as an ascetic, now becomes a kind of magician, a man apart, the potential holder of a universal key to all physical phenomena, thus endowed with a potentially omnipotent knowledge. This brings us back to an issue we have already raised: it is only in a simple world (especially in the world of classical science, where complexity merely veils a fundamental simplicity) that a form of knowledge that provides a universal key can exist.

One of the problems of our time is to overcome attitudes that tend to justify and reinforce the isolation of the scientific community. We must open new channels of communication between science and society. It is in this spirit that this book has been written. We all know that man is altering his natural environment on an unprecedented scale. As Serge Moscovici puts it, he is creating a "new nature."[21] But to understand this man-made world, we need a science that is not merely a tool submissive to external interests, nor a cancerous tumor irresponsibly growing on a substrate society.

Two thousand years ago Chuang Tsu wrote:

> How [ceaselessly] Heaven revolves! How [constantly] Earth abides at rest! Do the Sun and the Moon contend about their respective places? Is there someone presiding over and directing those things? Who binds and connects them together? Who causes and maintains them without trouble or exertion? Or is there perhaps some secret mechanism in consequence of which they cannot but be as they are?[22]

We believe that we are heading toward a new synthesis, a new naturalism. Perhaps we will eventually be able to combine the Western tradition, with its emphasis on experimentation and quantitative formulations, with a tradition such as the Chinese one, with its view of a spontaneous, self-organizing world. Toward the beginning of this Introduction, we cited Jacques Monod. His conclusion was: "The ancient alliance has been destroyed; man knows at last that he is alone in the universe's indifferent immensity out of which he emerged only by chance."[23] Perhaps Monod was right. The ancient alliance has been shattered. Our role is not to lament the past. It is to try to discover in the midst of the extraordinary diversity of the sciences some unifying thread. Each great period of science has led to some model of nature. For classical science it was the clock; for nineteenth-century science, the period of the Industrial Revolution, it was an engine running down. What will be the symbol for us? What we have in mind may perhaps be expressed best by a reference to sculpture, from Indian or pre-Columbian art to our time. In some of the most beautiful manifestations of sculpture, be it in the dancing Shiva or in the miniature temples of Guerrero, there appears very clearly the search for a junction between stillness and motion, time arrested and time passing. We believe that this confrontation will give our period its uniqueness.

BOOK ONE

THE DELUSION OF THE UNIVERSAL

CHAPTER I

THE TRIUMPH OF REASON

The New Moses

Nature and Nature's laws lay hid in night:
God said, let Newton be! and all was light.
—Alexander Pope,
Proposed Epitaph for Isaac Newton,
who died in 1727

There is nothing odd in the dramatic tone employed by Pope. In the eyes of eighteenth-century England, Newton was the "new Moses" who had been shown the "tables of the law." Poets, architects, and sculptors joined to propose monuments; a whole nation assembled to celebrate this unique event: a man had discovered the language that nature speaks—and obeys.

Nature compelled, his piercing Mind obeys,
And gladly shows him all her secret Ways;
'Gainst Mathematicks she has no Defence,
And yields t'experimental Consequence.[1]

Ethics and politics drew upon the Newtonian episode for material on which to "ground" their arguments. Thus Desaguliers transposed the meaning of the new natural order into a political lesson: a constitutional monarchy is the best possible system of government, since the King, like the Sun, has his power limited by it.

Like Ministers attending ev'ry Glance
Six Worlds sweep round his Throne in Mystick Dance.

He turns their Motion from his Devious Course,
And bends their Orbits by Attractive Force;
His Pow'r coerc'd by Laws, still leave them free,
Directs, but not Destroys, their Liberty;[2]

Although he himself did not encroach upon the domain of the moral sciences, Newton had no hesitation regarding the universal nature of the laws set out in his *Principia*. Nature is "very consonant and conformable to herself," he asserts in the celebrated Question 31 of his *Opticks*—and this strong and elliptical statement conceals a vast claim: combustion, fermentation, heat, cohesion, magnetism . . . there is no natural process which would not be produced by these active forces—attractions and repulsions—that govern both the motion of the stars and that of freely falling bodies.

Already a national hero before his death, nearly a century later Newton was to become, mainly through the powerful influence exerted by Laplace, the symbol of the scientific revolution in Europe. Astronomers scanned a sky ruled by mathematics. The Newtonian system succeeded in overcoming all obstacles. Furthermore, it opened the way to mathematical methods by which apparent deviations could be accounted for and even be used to infer the existence of a hitherto unknown planet. The prediction of the existence of the planet Neptune was the consecration of the prophetic power inherent in the Newtonian vision.

At the dawn of the nineteenth century, Newton's name tended to signify anything that claimed exemplarity. However, conflicting interpretations of his method are given. Some saw it as providing a blueprint for quantitative experimentation expressible in mathematics. For them, chemistry found its Newton in Lavoisier, who pioneered the systematic use of the balance. This was indeed a decisive step in the definition of a quantitative chemistry that took mass conservation as its Ariadne's thread. According to others, the Newtonian strategy consisted in isolating some central, specific fact and then using it as the basis for all further deductions concerning a given set of phenomena. In this perspective Newton's genius was located in his pragmatism. He did not try to explain gravitation; he took it as a fact. Similarly, each discipline should then take some central unexplained fact as its starting point. Physicians thus felt that they were authorized by Newton to refashion the vitalist conception and to speak of a "vital force" *sui generis,* the use of which would give the description of living phenomena a hoped-for systematic consistency. This is the same role that affinity, taken as the specifically chemical force of interaction, was called upon to play.

Some "true Newtonians" took exception to this proliferation of forces and reasserted the universality of the explanatory power of gravitation. But it was too late. The term *Newtonian* was now applied to everything that dealt with a system of laws, with equilibrium, or even to all situations in which natural order on one side and moral, social, and political order on the other could be expressed in terms of an all-embracing harmony. Romantic philosophers even discovered in the Newtonian universe an enchanted world animated by natural forces. More "orthodox" physicists saw in it a mechanical world governed by mathematics. For the positivists it meant the success of a procedure, a recipe to be identified with the very definition of science.[3]

The rest is literature—often Newtonian literature: the harmony that reigns in the society of stars, the elective affinities and hostilities giving rise to the "social life" of chemical compounds appear as processes that can be transposed into the world of human society. No wonder that this period appears as the Golden Age of Classical Science.

Today Newtonian science still occupies a unique position. Some of the basic concepts it introduced represent a definitive acquisition that has survived all the mutations science has since undergone. However, today we know that the Golden Age of Classical Science is gone, and with it also the conviction that Newtonian rationality, even with its various conflicting interpretations, forms a suitable basis for our dialogue with nature.

A central subject of this book is that of the Newtonian triumph, the continual opening up of new fields of investigation that have extended Newtonian thought right down to the present day. It also deals with doubts and struggles that arose from this triumph. Today we are beginning to see more clearly the limits of Newtonian rationality. A more consistent conception of science and of nature seems to be emerging. This new conception paves the way for a new unity of knowledge and culture.

A Dehumanized World

> . . . May God us keep
> From single Vision and Newton's sleep!
> —William Blake,
> in a letter to Thomas Butts
> dated November 22, 1802

There is no better illustration of the instability of the cultural position of Newtonian science than the introduction to a UNESCO colloquium on the relationship between science and culture:

> For more than a century the sector of scientific activity has been growing to such an extent within the surrounding cultural space that it seems to be replacing the totality of the culture itself. Some believe that this is merely an illusion due to its high growth rate and that the lines of force of this culture will soon reassert themselves and bring science back into the service of man. Others consider that the recent triumph of science entitles it at last to rule over the whole of culture which, moreover, would deserve to go on being known as such only because it was transmitted through the scientific apparatus. Others again, appalled by the danger of man and society being manipulated if they come under the sway of science, perceive the spectre of cultural disaster looming in the distance.[4]

In this statement science appears as a cancer in the body of culture, a cancer whose proliferation threatens to destroy the whole of cultural life. The question is whether we can dominate science and control its development, or whether we shall be enslaved. In only one hundred fifty years, science has been downgraded from a source of inspiration for Western culture to a threat. Not only does it threaten man's material existence, but also, more subtly, it threatens to destroy the traditions and experiences that are most deeply rooted in our cultural life. It is not just the technological fallout of one or another scientific breakthrough that is being accused, but "the spirit of science" itself.

Whether the accusation refers to a global skepticism exuded by scientific culture or to specific conclusions reached through scientific theories, it is often asserted today that science is debasing our world. What for generations had been a source of joy and amazement withers at its touch. Everything it touches is dehumanized.

Oddly enough, the idea of a fatal disenchantment brought about by scientific progress is an idea held not only by the critics of science but often also by those who defend or glorify it. Thus, in his book *The Edge of Objectivity,* historian C. C. Gillispie expresses sympathy for those who criticize science and constantly endeavor to blunt the "cutting edge of objectivity":

> Indeed, the renewals of the subjective approach to nature make a pathetic theme. Its ruins lie strewn like good intentions all along the ground traversed by science, until it survives only in strange corners like Lysenkoism and anthroposophy, where nature is socialized or moralized. Such survivals are relics of the perpetual attempt to escape the consequences of western man's most characteristic and successful campaign, which must doom to conquer. So like any thrust in the face of the inevitable, romantic natural philosophy has induced every nuance of mood from desperation to heroism. At the ugliest, it is sentimental or vulgar hostility to intellect. At the noblest, it inspired Diderot's naturalistic and moralizing science, Goethe's personification of nature, the poetry of Wordsworth, and the philosophy of Alfred North Whitehead, or of any other who would find a place in science for our qualitative and aesthetic appreciation of nature. It is the science of those who would make botany of blossoms and meteorology of sunsets.[5]

Thus science leads to a tragic, metaphysical choice. Man has to choose between the reassuring but irrational temptation to seek in nature a guarantee of human values, or a sign pointing to a fundamental correlatedness, and fidelity to a rationality that isolates him in a silent world.

The echoes of another leitmotiv—domination—mingle with that of disenchantment. A disenchanted world is, at the same time, a world liable to control and manipulation. Any science that conceives of the world as being governed according to a universal theoretical plan that reduces its various riches to the drab applications of general laws thereby becomes an instrument of domination. And man, a stranger to the world, sets himself up as its master.

This disenchantment has taken various forms in recent decades. It is outside the aim of this book to study systematically the various forms of antiscience. In Chapter III we shall present a fuller reaction of Western thought to the surprising triumph of Newtonian rationality. Here let us only note that at present there is a shift of popular attitudes to nature associated with a widespread but in our opinion erroneous belief that there exists a fundamental antagonism between science and "naturalism." To illustrate at least some of the forms antiscientific criticism has taken in recent years, we have chosen three examples. First, Heidegger, whose philosophy holds a deep fascination for contemporary thought. We shall also refer to the criticisms stated by Arthur Koestler and by the great historian of science, Alexander Koyré.

Martin Heidegger directs his criticism against the very core of the scientific endeavor, which he sees as fundamentally related to a permanent aim, the domination of nature. Therefore Heidegger claims that scientific rationality is the final accomplishment of something that has been implicitly present since ancient Greece, namely, the will to dominate, which is at work in any rational discussion or enterprise, the violence lurking in all positive and communicable knowledge. Heidegger emphasizes what he calls the technological and scientific "framing" *(Gestell)*,[6] which leads to the general setting to work of the world and of men.

Thus Heidegger does not present a detailed analysis of any particular technological or scientific product or process. What he challenges is the essence of technology, the way each thing is *taken into account*. Each theory is part of the implementation of the master plan that makes up Western history. What we call a scientific "theory" implies, following Heidegger, a way of questioning things by which they are reduced to enslavement. The scientist, like the technologist, is a toy in the hands of the will to power disguised as thirst for knowledge; his very approach to things subjects them to systematic violence.

> Modern physics is not experimental physics because it uses experimental devices in its questioning of nature. Rather the reverse is true. Because physics, already as pure theory, requests nature to manifest itself in terms of predictable forces, it sets up the experiments precisely for the sole purpose of asking whether and how nature follows the scheme preconceived by science.[7]

Similarly, Heidegger is not concerned about the fact that industrial pollution, for example, has destroyed all animal life in the Rhine. What does concern him is that the river has been put to man's service.

> The hydroelectric plant is set into the current of the Rhine. It sets the Rhine to supplying its hydraulic pressure, which then sets the turbines turning. . . . The hydroelectric plant is not built into the Rhine river as was the old bridge that joined bank with bank for hundreds of years. Rather the river is dammed up into the power plant. What the river is now, namely, a water supplier, derives from out of the essence of the power station.[8]

The old bridge over the Rhine is valued not as a proof of soundly tested ability, of painstaking and accurate observation, but because it does not "use" the river.

Heidegger's criticisms, taking the very ideal of a positive, communicable knowledge as a threat, echo some themes of the antiscience movement to which we referred in the Introduction. But the idea of an indissociable link between science and the will to dominate also permeates some apparently very different assessments of our present-day situation. For instance, under the very suggestive title "The Coming of the

Golden Age,"[9] Gunther Stent states that science is now reaching its limits. We are close to a point of diminishing returns, where the questions we direct to things in order to master them become more and more complicated and devoid of interest. This marks the end of progress, but it is the opportunity for humanity to stop its frantic efforts, to end the age-old struggle against nature, and to accept a static and comfortable peace. We wish to show that the relative dissociation between the scientific knowledge of an object and the possibility of mastering it, far from marking the end of science, signals a host of new perspectives and problems. Scientific understanding of the world around us is just beginning. There is yet another idea of science that we feel is potentially just as detrimental, namely, the fascination with a mysterious science that, by paths of reasoning inaccessible to common mortals, will lead to results that can, in one fell swoop, challenge the meaning of basic concepts such as time, space, causality, mind, or matter. This kind of "mystery science," the results of which are imagined to be capable of shattering the framework of any traditional conception, has actually been encouraged by the successive "revelations" of relativity and quantum mechanics. It is certainly true that some of the most imaginative steps in the past, Einstein's interpretation of gravitation as a space curvature or Dirac's antiparticles, for example, have shaken some seemingly well-established conceptions. Thus there is a very delicate balance between the readiness to imagine that science can produce anything and a kind of down-to-earth realism. Today the balance is strongly shifting toward a revival of mysticism, be it in the press media or even in science itself, especially among cosmologists.[10] It has even been suggested by certain physicists and popularizers of science that mysterious relationships exist between parapsychology and quantum physics. Let us cite Koestler:

> We have heard a whole chorus of Nobel Laureates in physics informing us that matter is dead, causality is dead, determinism is dead. If that is so, let us give them a decent burial, with a requiem of electronic music. It is time for us to draw the lesson from twentieth-century post-mechanistic science, and to get out of the strait-

> jacket which nineteenth-century materialism imposed on our philosophical outlook. Paradoxically, had that outlook kept abreast with modern science itself, instead of lagging a century behind it, we would have been liberated from that strait-jacket long ago. . . . But once this is recognized, we might become more receptive to phenomena around us which one-sided emphasis on physical science has made us ignore; might feel the draught that is blowing through the chinks of the causal edifice; pay more attention to confluential events; include the paranormal phenomena in our concept of normality; and realise that we have been living in the "Country of the Blind."[11]

We do not wish to judge or condemn a priori. There may be in some of the apparently fantastic propositions we hear today some seed of new knowledge. Nevertheless, we believe that leaps into the unimaginable are far too simple escapes from the concrete complexity of our world. We do not believe we shall leave the "Country of the Blind" in a day, since conceptual blindness is not the main reason for the problems and contradictions our society has failed to solve.

Our disagreement with certain criticisms or distortions of science does not mean, however, that we wish to reject all criticisms. Let us take, for instance, the position of Alexander Koyré, who has made outstanding contributions to the understanding of the development of modern science. In his study of the significance and implications of the Newtonian synthesis, Koyré wrote:

> Yet there is something for which Newton—or better to say not Newton alone, but modern science in general—can still be made responsible: it is the splitting of our world in two. I have been saying that modern science broke down the barriers that separated the heavens and the earth, and that it united and unified the universe. And that is true. But, as I have said, too, it did this by substituting for our world of quality and sense perception, the world in which we live, and love, and die, another world—the world of quantity, of reified geometry, a world in which, though there is a place for everything, there is

> no place for man. Thus the world of science—the real world—became estranged and utterly divorced from the world of life, which science has been unable to explain—not even to explain away by calling it "subjective."
>
> True, these worlds are everyday—and even more and more—connected by the *practice*. Yet for *theory* they are divided by an abyss.
>
> Two worlds: this means two truths. Or no truth at all.
>
> This is the tragedy of the modern mind which "solved the riddle of the universe," but only to replace it by another riddle: the riddle of itself.[12]

However, we hear in the conclusions of Koyré the same theme expressed by Pascal and Monod—this tragic feeling of estrangement. Koyré's criticism does not challenge scientific thinking but rather classical science based on the Newtonian perspective. We no longer have to settle for the previous dilemma of choosing between a science that reduces man to being a stranger in a disenchanted world and antiscientific, irrational protests. Koyré's criticism does not invoke the limits of a "strait-jacket" rationality but only the incapacity of classical science to deal with some fundamental aspects of the world in which we live.

Our position in this book is that the science described by Koyré is no longer our science. Not because we are concerned today with new, unimaginable objects, closer to magic than to logic, but because as scientists we are now beginning to find our way toward the complex processes forming the world with which we are most familiar, the natural world in which living creatures and their societies develop. Indeed, today we are beginning to go beyond what Koyré called "the world of quantity" into the world of "qualities" and thus of "becoming." This will be the main subject of Books One and Two. We believe it is precisely this transition to a new description that makes this moment in the history of science so exciting. Perhaps it is not an exaggeration to say that it is a period like the time of the Greek atomists or the Renaissance, periods in which a new view of nature was being born. But let us first return to Newtonian science, certainly one of the great moments of human history.

The Newtonian Synthesis

What lay behind the enthusiasm of Newton's contemporaries, their conviction that the secret of the universe, the truth about nature, had finally been revealed? Several lines of thought, probably present from the very beginning of humanity, converge in Newton's synthesis: first of all, science as a way of acting on our environment. Newtonian science is indeed an *active* science; one of its sources is the knowledge of the medieval craftsmen, the knowledge of the builders of machines. This science provides the means for systematically acting on the world, for predicting and modifying the course of natural processes, for conceiving devices that can harness and exploit the forces and material resources of nature.

In this sense, modern science is a continuation of the ageless efforts of man to organize and exploit the world in which he lives. We have very scanty knowledge about the early stages of this endeavor. However, it is possible, in retrospect, to assess the knowledge and skills required for the "Neolithic Revolution" to take place, when man gradually began to organize his natural and social environment, using new techniques to exploit nature and to organize his society. We still use, or have used until quite recently, Neolithic techniques—for example, animal and plant species either bred or selected, weaving, pottery, metalworking. Our social organization was for a long time based on the same techniques of writing, geometry, and arithmetic as those required to organize the hierarchically differentiated and structured social groups of the Neolithic city-states. Thus we cannot help acknowledging the continuity that exists between Neolithic techniques and the scientific and industrial revolutions.[13]

Modern science has thus extended this ancient endeavor, amplifying it and constantly speeding up its rhythm. Nevertheless, this does not exhaust the significance of science in the sense given to it by the Newtonian synthesis.

In addition to the various techniques used in a given society, we find a number of beliefs and myths that seek to understand man's place in nature. Like myths and cosmologies, science's endeavor is to *understand* the nature of the world, the way it is organized, and man's place in it.

From our standpoint it is quite irrelevant that the early speculations of the pre-Socratics appear to be adapted from the Hesiodic myth of creation—that is, the initial polarization of Heaven and Earth, the desire aroused by Eros, the birth of the first generations of gods to form the differentiated cosmic powers, discord and strife, alternating atrocities and vendettas, until stability is finally reached under the rule of Justice (*dikè*). What does matter is that, in the space of a few generations, the pre-Socratics collected, discussed, and criticized some of the concepts we are still trying to organize in order to understand the relation between being and becoming, or the appearance of order out of a hypothetically undifferentiated initial environment.

Where does the instability of the homogeneous come from? Why does it differentiate spontaneously? Why do things exist at all? Are they the fragile and mortal result of an injustice, a disequilibrium in the static equilibrium of forces between conflicting natural powers? Or do the forces that create and drive things exist autonomously—rival powers of love and hate leading to birth, growth, decline, and dispersion? Is change an illusion or is it, on the contrary, the unceasing struggle between opposites that constitutes things? Can qualitative change be reduced to the motion in a vacuum, of atoms differing only in their forms, or do atoms themselves consist of a multitude of qualitatively different germs, each unlike the others? And last, is the harmony of the world mathematical? Are numbers the key to nature?

The numerical regularities among sounds that were discovered by the Pythagoreans are still part of our present theories. The mathematical schemes worked out by the Greeks form the first body of abstract thought in European history—that is, a thought whose results are communicable and reproducible for all reasoning human beings. The Greeks achieved for the first time a form of deductive knowledge that contained a degree of certainty unaffected by convictions, expectations, or passions.

The most important aspect common to Greek thought and to modern science, which contrasts with the religious and mythical form of inquiry, is thus the emphasis on critical discussion and verification.[14]

Little is known about this pre-Socratic philosophy that grew up in the Ionian cities and the colonies of Magna Graecia. Thus we can only speculate about the relationships that might have existed between the development of theoretical and cosmological hypotheses and the crafts and technological activities that flourished in those cities. Tradition tells that as a result of a hostile religious and social reaction, philosophers were accused of atheism and were either exiled or put to death. This early "recall to order" may serve as a symbol of the importance of social factors in the origin, and above all the growth, of conceptual innovations. To understand the success of modern science we also have to explain why its founders were as a rule not unduly persecuted and their theoretical approach repressed in favor of a form of knowledge more consistent with social anticipations and convictions.

Be that as it may, from Plato and Aristotle onward, the limits were set, and thought was channeled in socially acceptable directions. In particular, the distinction between *theoretical thinking* and *technological activity* was established. The words we still use today—machine, mechanical, engineer—have a similar meaning. They do not refer to rational knowledge but to cunning and expediency. The idea was not to learn about natural processes in order to utilize them more effectively, but to deceive nature, to "machinate" against it—that is, to work wonders and create effects extraneous to the "natural order" of things. The fields of practical manipulation and that of the rational understanding of nature were thus rigidly separated. Archimedes' status is merely that of an engineer; his mathematical analysis of the equilibrium of machines is not considered to be applicable to the world of nature, at least within the framework of traditional physics. In contrast, the Newtonian synthesis expresses a systematic alliance between manipulation and theoretical understanding.

There is a third important element that found its expression in the Newtonian revolution. There is a striking contrast, which each of us has probably experienced, between the quiet world of the stars and planets and the ephemeral, turbulent world around us. As Mircea Eliade has emphasized, in many ancient civilizations there is a separation between profane space and sacred space, a division of the world into an ordinary space that is subject to chance and degradation and a sacred one that is meaningful, independent of contingency and history. This was the very contrast Aristotle established between the world of the stars and our sublunar world. This contrast is crucial to the way in which Aristotle evaluated the possibility of a quantitative description of nature. Since the motion of the celestial bodies is not change but a "divine" state that is eternally the same, it may be described by means of mathematical idealizations. Mathematical precision and rigor are not relevant to the sublunar world. Imprecise natural processes can only be subjected to an approximate description.

In any case, for an Aristotelian it is more interesting to know why a process occurs than to describe *how* it occurs, or rather, these two aspects are indivisible. One of the main sources of Aristotle's thinking was the observation of embryonic growth, a highly organized process in which interlocking, although apparently independent, events participate in a process that seems to be part of some global plan. Like the developing embryo, the whole of Aristotelian nature is organized according to final causes. The purpose of all change, if it is in keeping with the nature of things, is to realize in each being the perfection of its intelligible essence. Thus this essence, which, in the case of living creatures, is at one and the same time their final, formal, and effective cause, is the key to the understanding of nature. In this sense the "birth of modern science," the clash between the Aristotelians and Galileo, is a clash between two forms of rationality.[15]

In Galileo's view the question of "why," so dear to the Aristotelians, was a very dangerous way of addressing nature, at least for a scientist. The Aristotelians, on the other hand, considered Galileo's attitude as a form of irrational fanaticism.

Thus, with the coming of the Newtonian system it was a new universality that triumphed, and its emergence unified what till then had appeared as divided.

The Experimental Dialogue

We have already emphasized one of the essential elements of modern science: the marriage between theory and practice, the blending of the desire to shape the world and the desire to understand it. For this to be possible, it was not enough, despite the empiricists' beliefs, merely to respect observed facts. On certain points, including even the description of mechanical motion, it was in fact Aristotelian physics that was more easily brought into contact with empirical facts. The experimental dialogue with nature discovered by modern science involves *activity* rather than passive observation. What must be done is to manipulate physical reality, to "stage" it in such a way that it conforms as closely as possible to a theoretical description. The phenomenon studied must be prepared and isolated until it approximates some *ideal situation* that may be physically unattainable but that conforms to the conceptual scheme adopted.

By way of example, let us take the description of a system of pulleys, a classic since the time of Archimedes, whose reasoning has been extended by modern scientists to cover all simple machines. It is astonishing to find that the modern explanation has eliminated, on the grounds that it is irrelevant, the very thing that Aristotelian physics set out to explain, namely, the fact that, using a typical image, a stone "resists" a horse's efforts to pull it and that this resistance can be "overcome" by applying traction through a system of pulleys. Nature, according to Galileo, never gives anything away, never does something for nothing, and can never be tricked; it is absurd to think that by cunning or by using some stratagem we can make it perform extra work.[16] Since the work the horse is able to perform is the same with or without the pulleys, the effect produced *must* be the same. This then becomes the starting point for a mechanical explanation, which thus refers to an idealized world. In this world the "new" effect—the stone finally set in motion—is of secondary importance; and the stone's resistance is described only qualitatively, in terms of friction and heating. Instead, what is described accurately is the ideal situation, in which a relationship of equivalence links the cause, the work done by the horse, to the effect, the motion of the stone. In this ideal world, *the horse can, in any case, shift the stone,* and the system of pulleys has the sole effect of modifying the way the pulling efforts are transmitted; instead of moving the stone over a distance L, equal to the distance it travels while pulling the rope, the horse only moves it over a distance L/n, where n depends on the number of pulleys. Like all simple machines, the pulleys form a passive device that can only transmit motion without producing it.

The experimental dialogue thus corresponds to a highly specific procedure. Nature is cross-examined through experimentation, as if in a court of law, in the name of a priori principles. Nature's answers are recorded with the utmost accuracy, but relevance of those answers is assessed in terms of the very idealizations that guided the experiment. All the rest does not count as information, but is idle chatter, negligible secondary effects. It may well be that nature rejects the theoretical hypothesis in question. Nevertheless, the latter is still used as a standard against which to measure the implications and the significance of the response, whatever it may be. It is precisely this imperative way of questioning nature that Heidegger refers to in his argument against scientific rationality.

For us the experimental method is truly an *art*—that is, it is based on special skills and not on general rules. As such there are never any guarantees of success and one always remains at the mercy of triviality or poor judgment. No methodological principle can eliminate the risk, for instance, of persisting in a blind alley of inquiry. The experimental method is the art of choosing an interesting question and of scanning all the consequences of the theoretical framework thereby implied, all the ways nature could answer in the theoretical language chosen. Amid the concrete complexity of natural phenomena, one phenomenon has to be selected as the most likely to embody the theory's implications in an unambiguous way. This phenomenon will then be abstracted from its environment and "staged" to allow the theory to be tested in a reproducible and communicable way.

Although this experimental procedure was criticized right from the outset, ignored by the empiricists, and attacked by others on the grounds that it was a kind of torture, a way of putting nature on the rack, it survived all the modifications of the theoretical content of scientific descriptions and ultimately defined the new method of investigation introduced by modern science.

Experimental procedure can even become a tool for purely theoretical analysis. It is then a "thought experiment," the imagining of experimental situations governed entirely by theoretical principles, which permits the exploration of the consequences of these principles in a given situation. Such thought experiments played a crucial role in Galileo's work, and today they are at the center of investigations about the consequences of the conceptual upheavals in contemporary physics, namely, relativity and quantum mechanics. One of the most famous of such thought experiments is Einstein's famous train, from which an observer can measure the velocity of propagation of a ray of light emitted along an embankment, that is, moving at a velocity c in a reference system with respect to which the train is moving at a velocity v. According to classical reasoning, the observer on the train should attribute to the light, which is traveling in the same direction as he is, a velocity of $c - v$. However, this classical conclusion represents precisely the absurdity that the thought experiment was designed to expose. In relativity theory, the velocity of light appears as a *universal* constant of nature. Whatever inertial reference system is used, the velocity of light is always the same. And since then Einstein's train has gone on exploring the physical consequences of this fundamental change.

The experimental method is central to the dialogue with nature established by modern science. Nature questioned in this way is, of course, simplified and occasionally mutilated. This does not deprive it of its capacity to refute most of the hypotheses we can imagine. Einstein used to say that nature says "no" to most of the questions it is asked, and occasionally "perhaps." The scientist does not do as he pleases, and he cannot force nature to say only what he wants to hear. He cannot, at least in the long run, project upon it his most cherished desires and expectations. He actually runs a greater risk and plays a more dangerous game the better his tactics succeed in encircling nature, in setting it more squarely with its back to the wall.[17] Moreover, it is true that, whether the answer is "yes" or "no," it will be expressed in the same theoretical language as the question. However, this language, too, develops according to a complex historical process involving nature's replies in the past and its relations with other theoretical languages. In addition, new questions arise corresponding to the changing interests of each period. This sets up a complex relationship between the specific rules of the scientific game—particularly the experimental method of reasoning with nature, which places the greatest constraint on the game—and a cultural network to which, sometimes unwittingly, the scientist belongs.

We believe that the experimental dialogue is an irreversible acquisition of human culture. It actually provides a guarantee that when nature is explored by man it is treated as an *independent* being. It forms the basis of the communicable and reproducible nature of scientific results. However partially nature is allowed to speak, once it has expressed itself, there is no further dissent: nature never lies.

The Myth at the Origin of Science

The dialogue between man and nature was accurately perceived by the founders of modern science as a basic step toward the intelligibility of nature. But their ambitions went even farther. Galileo, and those who came after him, conceived of science as being capable of discovering *global* truths about nature. Nature not only would be written in a mathematical language that can be deciphered by experimentation, but there would actually exist only one such language. Following this basic conviction, the world is seen as homogeneous, and local experimentation can reveal global truth. The simplest phenomena studied by science can thus be interpreted as the key to understanding nature as a whole; the complexity of the latter is only apparent, and its diversity can be explained in terms of the universal truth embodied, in Galileo's case, in the mathematical laws of motion.

This conviction has survived centuries. In an excellent set of lectures presented on the BBC several years ago, Richard Feynman[18] compared nature to a huge chess game. The complexity is only apparent; each move follows simple rules. In its early days, modern science quite possibly needed this conviction of being able to reach global truth. Such a conviction added an immense value to the experimental method and, to a certain extent, inspired it. Perhaps a revolutionary conception of the world, one as all-embracing as the "biological" conception of the Aristotelian world, was necessary to throw off the yoke of tradition, to give the champions of experimentation a strength of conviction and a power of argument that enabled them to hold their own against the previous forms of rationalism. Perhaps a metaphysical conviction was needed to transmute the craftsman's and machine builder's knowledge into a new method for the rational exploration of nature. We may also wonder what the implications of the existence of this kind of "mythical" conviction are for explaining the way modern science's first developments were accepted in the social context. On this highly controversial issue, we shall restrict ourselves to a few remarks of a quite general nature for the sole purpose of pinpointing the problem—that is, the problem of a science whose advance has been felt by some as the triumph of reason, but by others as a disillusionment, as the painful discovery of the robotlike stupidity of nature.

It seems hard to deny the fundamental importance of social and economic factors—particularly the development of craftsmen's techniques in the monasteries, where the residual knowledge of a destroyed world was preserved, and later in the bustling merchant cities—in the birth of experimental science, which is a systematized form of part of the craftsmen's knowledge.

Moreover, a comparative analysis such as Needham's[19] exposes the decisive importance of social structures at the close of the Middle Ages. Not only was the class of craftsmen and potential technical innovators not held in contempt, as it was in ancient Greece, but, like the craftsmen, the intellectuals were, in the main, independent of the authorities. They were free entrepreneurs, craftsmen-inventors in search of patronage, who tended to look for novelty and to exploit all the opportunities it afforded, however dangerous they may have been for the social order. On the other hand, as Needham points out, Chinese men of science were officials, bound to observe the rules of the bureaucracy. They formed an integral part of the state, whose primary objective was to keep law and order. The compass, the printing press, and gunpowder, all of which were to contribute to undermining the foundations of medieval society and to project Europe into the modern era, were discovered much earlier in China but had a much less destabilizing effect on its society. The enterprising European merchant society appears in contrast as particularly well suited to stimulate and sustain the dynamic and innovative growth of modern science in its early stages.

However, the question remains. We know that the builders of machines used mathematical concepts—gear ratios, the displacements of the various working parts, and the geometry of their relative motions. But why was mathematization not restricted to machines? Why was natural motion conceived of in the image of a rationalized machine? This question may also be asked in connection with the clock, one of the triumphs of medieval craftsmanship that was soon to set the rhythm of life in the larger medieval towns. Why did the clock almost immediately become the very symbol of world order? In this last question lies perhaps some elements of an answer. A watch is a *contrivance* governed by a rationality that lies outside itself, by a plan that is blindly executed by its inner workings. The clock world is a metaphor suggestive of God the Watchmaker, the rational master of a robotlike nature. At the origin of modern science, a "resonance" appears to have been set up between theological discourse and theoretical and experimental activity—a resonance that was no doubt likely to amplify and consolidate the claim that scientists were in the process of discovering the secret of the "great machine of the universe."

Of course, the term *resonance* covers an extremely complex problem. It is not our intention to state, nor are we in any position to affirm, that religious discourse in any way *determined* the birth of theoretical science, or of the "world view" that happened to develop in conjunction with experimental activity. By using the term *resonance*—that is, mutual amplification of two discourses—we have deliberately chosen an expression that does not assume whether it was theological discourse or the "scientific myth" that came first and triggered the other.

Let us note that to some philosophers the question of the "Christian origin" of Western science is not only the question of the stabilization of the concept of nature as an automaton, but also the question of some "essential" link between experimental science as such and Western civilization in its Hebraic and Greek components. For Alfred North Whitehead this link is situated at the level of instinctive conviction. Such a conviction was "needed" to inspire the "scientific faith" of the founders of modern science:

> I mean the inexpugnable belief that every detailed occurrence can be correlated with its antecedents in a perfectly definite manner, exemplifying general principles. Without this belief the incredible labours of scientists would be without hope. It is this instinctive conviction, vividly poised before the imagination, which is the motive power of research: that there is a secret, a secret which can be unveiled. How has this conviction been so vividly implanted in the European mind?
>
> When we compare this tone of thought in Europe with the attitude of other civilizations when left to themselves, there seems but one source for its origin. It must come from the medieval insistence on the rationality of God, conceived as with the personal energy of Jehovah and with the rationality of a Greek philosopher. Every detail was supervised and ordered: the search into nature could only result in the vindication of the faith in rationality. Remember that I am not talking of the explicit beliefs of a few individuals. What I mean is the impress on the European mind arising from the unquestioned faith of centuries. By this I mean the instinctive tone of thought and not a mere creed of words.[20]

We will not consider this matter further. It would be out of the question to "prove" that modern science could have originated only in Christian Europe. It is not even necessary to ask if the founders of modern science drew any real inspiration from theological arguments. Whether or not they were sincere, the important point is that those arguments made the speculations of modern science socially credible and acceptable, over a period of time varying from country to country. Religious references were still frequent in English scientific texts of the nineteenth century. Remarkably enough, in the present-day revival of interest in mysticism, the direction of the argument appears reversed. It is now science that appears to lend credibility to mystical affirmation.

The question we have confronted here obviously leads toward a multitude of problems in which theological and scientific issues are inextricably bound up with the "external" history of science, that is, the description of the relationship between the form and content of scientific knowledge on the one hand, and on the other, the use to which it is put in its social, economic, and institutional context. As we have already said, the only point we are presently interested in is the very particular character and implications of scientific discourse that was amplified by resonance with theological discourses.

Needham[21] tells of the irony with which Chinese men of letters of the eighteenth century greeted the Jesuits' announcement of the triumphs of modern science. The idea that nature was governed by simple, knowable laws appeared to them as a perfect example of anthropocentric foolishness. Needham believes that this "foolishness" has deep cultural roots. In order to illustrate the great differences between the Western and Chinese conceptions, he cites the animal trials held in the Middle Ages. On several occasions such freaks as a cock who supposedly laid eggs were solemnly condemned to death and burned for having infringed the laws of nature, which were equated with the laws of God. Needham explains how, in China, the same cock would, in all likelihood, merely have disappeared discreetly. It was not guilty of any crime, but its freakish behavior clashed with natural and social harmony. The governor of the province or even the emperor might find himself in a delicate situation if the misbehavior of the cock became known. Needham comments that, according to a philosophic conception dominant in China, the cosmos is in spontaneous harmony and the regularity of phenomena is not due to any external authority. On the contrary, this harmony in nature, society, and the heavens originates from the equilibrium among these processes. Stable and interdependent, they resonate with each other in a kind of nonconcerted harmony. If any law were involved, it would be a law that no one, neither God nor man, had ever conceived of. Such a law would also have to be expressed in a language undecipherable by man and not be a law established by a creator conceived in our own image.

Needham concludes by asking the following question:

> In the outlook of modern science there is, of course, no residue of the notions of command and duty in the "Laws" of Nature. They are now thought of as statistical regularities, valid only in given times and places, descriptions not prescriptions, as Karl Pearson put it in a famous chapter. The exact degree of subjectivity in the formulations of scientific law has been hotly debated during the whole period from Mach to Eddington, and such questions cannot be followed further here. The problem is whether the recognition of such statistical regularities and their mathematical expression could have been reached by any other road than that which Western science actually travelled. Was perhaps the state of mind in which an egg-laying cock could be prosecuted at law necessary in a culture which should later have the property of producing a Kepler?[22]

It must now be stressed that scientific discourse is in no way a mere transposition of traditional religious views. Obviously the world described by classical physics is not the world of Genesis, in which God created light, heaven, earth, and the living species, the world where Providence has never ceased to act, spurring man on toward a history where his salvation is at stake. The world of classical physics is an atemporal world which, if created, must have been created in one fell swoop, somewhat as an engineer creates a robot before letting it function alone. In this sense, physics has indeed developed in opposition to both religion and the traditional philosophies. And yet we know that the Christian God was actually called upon to provide a basis for the world's intelligibility. In fact, one can speak here of a kind of "convergence" between the interests of theologians, who held that the world had to acknowledge God's omnipotence by its total submission to Him, and of physicists seeking a world of mathematizable processes.

In any case, the Aristotelian world destroyed by modern science was unacceptable to both these theologians and physicists. This ordered, harmonious, hierarchical, and rational world was too independent, the beings inhabiting it too powerful and active, and their subservience to the absolute sovereign too suspect and limited for the needs of many theologians.[23] On the other hand, it was too complex and qualitatively differentiated to be mathematized.

The "mechanized" nature of modern science, created and ruled according to a plan that totally dominates it, but of which it is unaware, glorifies its creator, and was thus admirably suited to the needs of both theologians and the physicists. Although Leibniz had endeavored to demonstrate that mathematization is compatible with a world that can display active and qualitatively differentiated behavior, scientists and theologians joined forces to describe nature as a mindless, passive mechanics that was basically alien to freedom and the purposes of the human mind. "A dull affair, soundless, scentless, colourless, merely the hurrying of matter, endless, meaningless,"[24] as Whitehead observes. This Christian nature, stripped of any property that permits man to identify himself with the ancient harmony of natural "becoming," leaving man alone, face to face with God, thus converged with the nature that a single language, and not the thousand mathematical voices heard by Leibniz, was sufficient to describe.

Theology may also help comment on man's odd position when he laboriously deciphers the laws governing the world. Man is emphatically not part of the nature he objectively describes; he dominates it from the outside. Indeed, for Galileo, the human soul, created in God's image, is capable of grasping the intelligible truths underlying the plan of creation. It can thus gradually approach a knowledge of the world that God himself possessed intuitively, fully, and instantaneously.[25]

Unlike the ancient atomists, who were persecuted on the grounds of atheism, and unlike Leibniz, who was sometimes suspected of denying the existence of grace or of human freedom, modern scientists have managed to come up with a culturally acceptable definition of their enterprise. The human mind, incorporated in a body subject to the laws of nature, can, by means of experimental devices, obtain access to the vantage point from which God himself surveys the world, to the divine plan of which this world is a tangible expression. Nevertheless, the mind itself remains outside the results of its achievement. The scientist may describe as secondary qualities, not part of nature but projected onto it by the mind, everything that goes to make up the texture of nature, such as its perfumes and its colors. The debasement of nature is parallel to the glorification of all that eludes it, God and man.

The Limits of Classical Science

We have tried to describe the unique historical situation in which scientific practice and metaphysical conviction were closely coupled. Galileo and those who came after him raised the same problems as the medieval builders but broke away from their empirical knowledge to assert, with the help of God, the simplicity of the world and the universality of the language the experimental method postulated and deciphered. In this way, the basic myth underlying modern science can be seen as a product of the peculiar complex which, at the close of the Middle Ages, set up conditions of resonance and reciprocal amplification among economic, political, social, religious, philosophic, and technical factors. However, the rapid decomposition of this complex left classical science stranded and isolated in a transformed culture.

Classical science was born in a culture dominated by the alliance between *man,* situated midway between the divine order and the natural order, and *God,* the rational and intelligible legislator, the sovereign architect we have conceived in our own image. It has outlived this moment of cultural consonance that entitled philosophers and theologians to engage in science and that entitled scientists to decipher and express opinions on the divine wisdom and power at work in creation. With the support of religion and philosophy, scientists had come to believe their enterprise was self-sufficient, that it exhausted the possibilities of a rational approach to natural phenomena. The relationship between scientific description and natural philosophy did not, in this sense, have to be justified. It could be seen as self-evident that science and philosophy were convergent and that science was discovering the principles of an authentic natural philosophy. But, oddly enough, the self-sufficiency experienced by scientists was to outlive the departure of the medieval God and the withdrawal of the epistemological guarantee offered by theology. The originally bold bet had become the triumphant science of the eighteenth century,[26] the science that discovered the laws governing the motion of celestial and earthly bodies, a science that d'Alembert and Euler incorporated into a complete and consistent system and whose history was defined by Lagrange as a logical achievement tending toward perfection. It was the science honored by the Academies founded by absolute monarchs such as Louis XIV, Frederick II, and Catherine the Great,[27] the science that made Newton a national hero. In other words, it was a *successful* science, convinced that it had *proved* that nature is transparent. *"Je n'ai pas besoin de cette hypothèse"* was Laplace's reply to Napoleon, who had asked him God's place in his world system.

The dualist implications of modern science were to survive as well as its claims. For the science of Laplace which, in many respects, is still the classical conception of science today, a description is objective to the extent to which the observer is excluded and the description itself is made from a point lying *de jure* outside the world, that is, from the divine viewpoint to which the human soul, created as it was in God's image, had access at the beginning. Thus classical science still aims at discovering the unique truth about the world, the one language that will decipher the whole of nature—today we would speak of the *fundamental level of description* from which everything in existence can be deduced.

On this essential point let us cite Einstein, who has translated into modern terms precisely what we may call the basic myth underlying modern science:

> What place does the theoretical physicist's picture of the world occupy among all these possible pictures? It demands the highest possible standard of rigorous precision in the description of relations, such as only the use of mathematical language can give. In regard to his subject matter, on the other hand, the physicist has to limit himself very severely: he must content himself with describing the most simple events which can be brought within the domain of our experience; all events of a more complex order are beyond the power of the human intellect to reconstruct with the subtle accuracy and logical perfection which the theoretical physicist demands. Supreme purity, clarity, and certainty at the cost of completeness.

> But what can be the attraction of getting to know such a tiny section of nature thoroughly, while one leaves everything subtler and more complex shyly and timidly alone? Does the product of such a modest effort deserve to be called by the proud name of a theory of the universe?
>
> In my belief the name is justified; for the general laws on which the structure of theoretical physics is based claim to be valid for any natural phenomenon whatsoever. With them, it ought to be possible to arrive at the description, that is to say, the theory, of every natural process, including life, by means of pure deduction, if that process of deduction were not far beyond the capacity of the human intellect. The physicist's renunciation of completeness for his cosmos is therefore not a matter of fundamental principle.[28]

For some time there were those who persisted in the illusion that attraction in the form in which it is expressed in the law of gravitation would justify attributing an intrinsic animation to nature and that if it were generalized it would explain the origins of increasingly specific forms of activity, including even the interactions that compose human society. But this hope was rapidly crushed, at least partly as a consequence of the demands created by the political, economic, and institutional setting where science developed. We shall not examine this aspect of the problem, important though it is. Our point here is to emphasize that this very failure seemed to establish the consistency of the classical view and to prove that what had once been an inspiring conviction was a sad truth. In fact, the only interpretation apparently capable of rivaling this interpretation of science was henceforth the positivistic refusal of the very project of understanding the world. For example, Ernst Mach, the influential philosopher-scientist whose ideas had a great impact on the young Einstein, defined the task of scientific knowledge as arranging experience in as economical an order as possible. Science has no other meaningful goal than the simplest and most economical abstract expression of facts:

> Here we have a clue which strips science of all its mystery, and shows us what its power really is. With respect

> to specific results it yields us nothing that we could not reach in a sufficiently long time without methods. . . . Just as a single human being, restricted wholly to the fruits of his own labor, could never amass a fortune, but on the contrary the accumulation of the labor of many men in the hands of one is the foundation of wealth and power, so, also, no knowledge worthy of the name can be gathered up in a single human mind limited to the span of a human life and gifted only with finite powers, except by the most exquisite economy of thought and by the careful amassment of the economically ordered experience of thousands of co-workers.[29]

Thus science is useful because it leads to economy of thought. There may be some element of truth in such a statement, but does it tell the whole story? How far we have come from Newton, Leibniz, and the other founders of Western science, whose ambition was to provide an intelligible frame to the physical universe! Here science leads to interesting rules of action, but no more.

This brings us back to our starting point, to the idea that it is *classical* science, considered for a certain period of time as the very symbol of cultural unity, and not science as such that led to the cultural crisis we have described. Scientists found themselves reduced to a blind oscillation between the thunderings of "scientific myth" and the silence of "scientific seriousness," between affirming the absolute and global nature of scientific truth and retreating into a conception of scientific theory as a pragmatic recipe for effective intervention in natural processes.

As we have already stated, we subscribe to the view that classical science has now reached its limit. One aspect of this transformation is the discovery of the limitations of classical concepts that imply that a knowledge of the world "as it is" was possible. The omniscient beings, Laplace's or Maxwell's demon, or Einstein's God, beings that play such an important role in scientific reasoning, embody the kinds of extrapolation physicists thought they were allowed to make. As randomness, complexity, and irreversibility enter into physics as objects of positive knowledge, we are moving away from this rather naïve assumption of a direct connection between our description of the world and the world itself. Objectivity in theoretical physics takes on a more subtle meaning.

This evolution was forced upon us by unexpected supplemental discoveries that have shown that the existence of universal constants, such as the velocity of light, limit our power to manipulate nature. (We shall discuss this unexpected situation in Chapter VII.) As a result, physicists had to introduce new mathematical tools that make the relation between perception and interpretation more complex. Whatever reality may mean, it always corresponds to an active intellectual construction. The descriptions presented by science can no longer be disentangled from our questioning activity and therefore can no longer be attributed to some omniscient being.

On the eve of the Newtonian synthesis, John Donne lamented the passing of the Aristotelian cosmos destroyed by Copernicus:

> And new Philosophy calls all in doubt,
> The Element of fire is quite put out,
> The Sun is lost, and th'earth, and no man's wit
> Can well direct him where to look for it.
> And freely men confess that this world's spent,
> When in the Planets and the Firmament,
> They seek so many new, then they see that this
> Is crumbled out again to his Atomies
> 'Tis all in Pieces, all coherence gone.[30]

The scattered bricks and stones of our present culture seem, as in Donne's time, capable of being rebuilt into a new "coherence." Classical science, the mythical science of a simple, passive world, belongs to the past, killed not by philosophical criticism or empiricist resignation but by the internal development of science itself.

CHAPTER II

THE IDENTIFICATION OF THE REAL

Newton's Laws

We shall now take a closer look at the mechanistic world view as it emerged from the work of Galileo, Newton, and their successors. We wish to describe its strong points, the aspects of nature it has succeeded in clarifying, but we also want to expose its limitations.

Ever since Galileo, one of the central problems of physics has been the description of acceleration. The surprising feature was that the change undergone by the state of motion of a body could be formulated in simple mathematical terms. This seems almost trivial to us today. Still, we should remember that Chinese science, so successful in many areas, did not produce a quantitative formulation of the laws of motion. Galileo discovered that we do not need to ask for the *cause* of a state of motion if the motion is uniform, any more than it is necessary to ask the reason for a state of rest. Both motion and rest remain indefinitely stable unless something happens to upset them. The central problem is the *change* from rest to motion, and from motion to rest, as well as, more generally, all changes of velocity. How do these changes occur? The formulation of the Newtonian laws of motion made use of two converging developments: one in physics, Kepler's laws for planetary motion and Galileo's laws for falling bodies, and the other in mathematics, the formulation of differential or "infinitesimal" calculus.

How can a continuously varying speed be defined? How can we describe the instantaneous changes in the various quantities, such as position, velocity, and acceleration? How can we describe the state of a body at any given instant? To answer these questions, mathematicians have introduced the concept of infinitesimal quantities. An infinitesimal quantity is the result of a *limiting process;* it is typically the variation in a quantity occurring between two successive instants when the time elapsing between these instants tends toward zero. In this way the change is broken up into an infinite series of infinitely small changes.

At each instant the state of a moving body can be defined by its position r, by its velocity v, which expresses its "instantaneous tendency" to modify this position, and by its acceleration a, again its "instantaneous tendency," but now to modify its velocity. Instantaneous velocities and accelerations are limiting quantities that measure the ratio between two infinitesimal quantities: the variation of r (or v) during a temporal interval Δt, and this interval Δt when Δt tends to zero. Such quantities are "derivatives with respect to time," and since Leibniz they have been written as $v = dr/dt$ and $a = dv/dt$. Therefore, acceleration, the derivative of a derivative, $a = d^2r/dt^2$, becomes a "second derivative." The problem on which Newtonian physics concentrates is the calculation of this second derivative, that is, of the acceleration undergone at each instant by the points that form a system. The motion of each of these points over a finite interval of time can then be calculated by *integration,* by adding up the infinitesimal velocity changes occurring during this interval. The simplest case is when a is constant (for example, for a freely falling body a is the gravitational constant g). Generally speaking, acceleration itself varies in time, and the physicist's task is to determine precisely the nature of this variation.

In Newtonian language, to study acceleration means to determine the various "forces" acting on the points in the system under examination. Newton's second law, $F = ma$, states that the force applied at any point is proportional to the acceleration it produces. In the case of a system of material points, the problem is more complicated, since the forces acting on a given body are determined at each instant by the relative distances between the bodies of the system, and thus vary at each instant as a result of the motion they themselves produce.

A problem in dynamics is expressed in the form of a set of "differential" equations. The instantaneous state of each of the bodies in a system is described as a point and defined by means of its position as well as by its velocity and acceleration, that is, by the first and second derivatives of the position. At each instant, a set of forces, which is a function of the distance between the points in the system (a function of r), gives a precise acceleration to each point; the accelerations then bring about changes in the distances separating these points and therefore in the set of forces acting at the following instant.

While the differential equations *set up* the dynamics problem, their "integration" represents the *solution* of this problem. It leads to the calculation of the *trajectories*, $r(t)$. These trajectories contain all the information acknowledged as relevant by dynamics; it provides a complete description of the dynamic system.

The description therefore implies two elements: the *positions and velocities* of each of the points at *one instant,* often called the "initial instant," and the equations of motion that relate the dynamic forces to the accelerations. The integration of the dynamic equations starting from the "initial state" unfold the succession of states, that is, the set of trajectories of its constitutive bodies.

The triumph of Newtonian science is the discovery that a single force, gravity, determines both the motion of planets and comets in the sky and the motions of bodies falling toward the earth. Whatever pair of material bodies is considered, the Newtonian system implies that they are linked by the same force of attraction. Newtonian dynamics thus appears to be doubly universal. The definition of the law of gravity that describes how masses tend to approach one another contains no reference to any scale of phenomena. It can be applied equally well to the motion of atoms, of planets, or of the stars in a galaxy. Every body, whatever its size, has a mass and acts as a source of the Newtonian forces of interaction.

Since gravitational forces connect any two bodies (for two bodies of mass m and m' and separated by a distance r, the gravitational force is kmm'/r^2, where k is the Newtonian gravitational constant equal to 6.67 $cm^3g^{-1}sec^{-2}$ 10^{-8}), the only true dynamic system is the universe as a whole. Any local dynamic system, such as our planetary system, can only be defined approximately, by neglecting forces that are small in comparison to those whose effect is being considered.

It must be emphasized that whatever the dynamic system chosen, the laws of motion can always be expressed in the form $F = ma$. Other types of forces apart from those due to gravity may be discovered (and actually have been discovered—for instance, electric forces of attraction and repulsion) and would thereby modify the empirical content of the laws of motion. They would not, however, modify the form of those laws. In the world of dynamics, change is identified with acceleration or deceleration. The integration of the laws of motion leads to the trajectories that the particles follow. Therefore the laws of change, of time's impact on nature, are expressed in terms of the characteristics of trajectories.

The basic characteristics of trajectories are *lawfulness, determinism,* and *reversibility.* We have seen that in order to calculate a trajectory we need, in addition to our knowledge of the laws of motion, an empirical definition of a single instantaneous state of the system. The general law then deduces from this "initial state" the series of states the system passes through as time progresses, just as logic deduces a conclusion from basic premises. The remarkable feature is that once the forces are known, any single state is sufficient to define the system completely, not only its future but also its past. At each instant, therefore, everything is given. Dynamics defines all states as equivalent: each of them allows all the others to be calculated along with the trajectory which connects all states, be they in the past or the future.

"Everything is given." This conclusion of classical dynamics, which Bergson repeatedly emphasized, characterizes the reality that dynamics describes. Everything is given, but everything is also possible. A being who has the power to control a dynamic system may calculate the right initial state in such a way that the system "spontaneously" reaches any chosen state at some chosen time. The generality of dynamic laws is matched by the arbitrariness of the initial conditions.

The *reversibility* of a dynamic trajectory was explicitly stated by all the founders of dynamics. For instance, when Galileo or Huyghens described the implications of the equivalence between cause and effect postulated as the basis of their mathematization of motion, they staged thought experiments such as an elastic ball bouncing on the ground. As the result of its instantaneous velocity inversion, such a body would return to its initial position. Dynamics assigns this property of reversibility to all dynamic changes. This early "thought experiment" illustrates a general mathematical property of dynamic equations. The structure of these equations implies that if the velocities of all the points of a system are reversed, the system will go "backward in time." The system would retrace all the states it went through during the previous change. Dynamics defines as mathematically equivalent changes such as $t \to -t$, time inversion, and $v \to -v$, velocity reversal. What one dynamic change has achieved, another change, defined by velocity inversion, can undo, and in this way exactly restore the original conditions.

This property of reversibility in dynamics leads, however, to a difficulty whose full significance was realized only with the introduction of quantum mechanics. Manipulation and measurement are essentially irreversible. *Active* science is thus, by definition, extraneous to the idealized, reversible world it is describing. From a more general point of view, reversibility may be taken as the very symbol of the "strangeness" of the world described by dynamics. Everyone is familiar with the absurd effects produced by projecting a film backward—the sight of a match being regenerated by its flame, broken ink pots that reassemble and return to a tabletop after the ink has poured back into them, branches that grow young again and turn into fresh shoots. In the world of classical dynamics, such events are considered to be just as likely as the normal ones.

We are so accustomed to the laws of classical dynamics that are taught to us early in school that we often fail to sense the boldness of the assumptions on which they are based. A world in which all trajectories are reversible is a strange world indeed. Another astonishing assumption is that of the complete independence of initial conditions from the laws of motion. It is possible to take a stone and throw it with some initial velocity limited only by one's physical strength, but what about a complex system such as a gas formed by many particles? It is obvious that we can no longer impose arbitrary initial conditions. The initial conditions must be the outcome of the dynamic evolution itself. This is an important point to which we shall come back in the third part of this book. But whatever its limitations, today, three centuries later, we can only admire the logical coherence and the power of the methods discovered by the founders of classical dynamics.

Motion and Change

Aristotle made time the measure of change. But he was fully aware of the qualitative multiplicity of change in nature. Still there is only one type of change surviving in dynamics, one "process," and that is motion. The qualitative diversity of changes in nature is reduced to the study of the relative displacement of material bodies. Time is a parameter in terms of which these displacements may be described. In this way space and time are inextricably tied together in the world of classical dynamics. (Also see Chapter IX.)

It is interesting to compare dynamic change with the atomists' conception of change, which enjoyed considerable favor at the time Newton formulated his laws. Actually, it seems that not only Descartes, Gassendi, and d'Alembert, but even Newton himself believed that collisions between hard atoms were the ultimate, and perhaps the only, sources of changes of motion.[1] Nevertheless, the dynamic and the atomic descriptions differ radically. Indeed, the continuous nature of the acceleration described by the dynamic equations is in sharp contrast with the discontinuous, instantaneous collisions between hard particles. Newton had already noticed that, in contradiction to dynamics, an irreversible loss of motion is involved in each hard collision. The only reversible collision—that is, the only one in agreement with the laws of dynamics—is the "elastic," momentum-conserving collision. But how can the complex property of "elasticity" be applied to atoms that are supposed to be the fundamental elements of nature?

On the other hand, at a less technical level, the laws of dynamic motion seem to contradict the randomness generally attributed to collisions between atoms. The ancient philosophers had already pointed out that any natural process can be interpreted in many different ways in terms of the motion of and collisions between atoms. This was not a problem for the atomists, since their main aim was to describe a godless, lawless world in which man is free and can expect to receive neither punishment nor reward from any divine or natural order. But classical science was a science of engineers and astronomers, a science of action and prediction. Speculations based on hypothetical atoms could not satisfy its needs. In contrast, Newton's law provided a means of predicting and manipulating. Nature thus becomes law-abiding, docile, and predictable, instead of being chaotic, unruly, and stochastic. But what is the connection between the mortal, unstable world in which atoms unceasingly combine and separate, and the immutable world of dynamics governed by Newton's law, a single mathematical formula corresponding to an eternal truth unfolding toward a tautological future? In the twentieth century we are again witnessing the clash between lawfulness and random events, which, as Koyré has shown, had already tormented Descartes.[2] Ever since the end of the nineteenth century, with the kinetic theory of gases, the atomic chaos has reintegrated physics, and the problem of the relationship between dynamic law and statistical description has penetrated to the very core of physics. It is one of the key elements in the present renewal of dynamics (see Book III).

In the eighteenth century, however, this contradiction seemed to produce a deadlock. This may partly explain the skepticism of some eighteenth-century physicists regarding the significance of Newton's dynamic description. We have already noted that collisions may lead to a loss of motion. They thereby concluded that in such nonideal cases, "energy" is not conserved but is *irreversibly* dissipated (see Chapter IV, section 3). Therefore, the atomists could not help considering dynamics as an idealization of limited value. Continental physicists and mathematicians such as d'Alembert, Clairaut, and Lagrange resisted the seductive charms of Newtonianism for a long time.

Where do the roots of the Newtonian concept of change lie? It appears to be a synthesis[3] of the science of ideal machines, where motion is transmitted without collision or friction between parts already in contact, and the science of celestial bodies interacting at a distance. We have seen that it appears as the very antithesis of atomism, which is based on the concept of random collisions. Does this, however, vindicate the view of those who believe that Newtonian dynamics represents a rupture in the history of thinking, a revolutionary novelty? This is what positivist historians have claimed when they described how Newton escaped the spell of preconceived notions and had the courage to infer from the mathematical study of planetary motions and the laws of falling bodies the action of a "universal" force. We know that on the contrary the eighteenth-century rationalists emphasized the apparent similarity between his "mathematical" forces and traditional occult qualities. Fortunately, these critics did not know the strange story behind the Newtonian forces! For behind Newton's cautious declaration—"I frame no hypotheses"—concerning the nature of the forces lurked the passion of an alchemist.[4] We now know that, side by side with his mathematical studies, Newton had studied the ancient alchemists for thirty years and had carried out painstaking laboratory experiments on ways of achieving the master work, the synthesis of gold.

Recently some historians have gone so far as to propose that the Newtonian synthesis of heaven and earth was the achievement of a chemist, not an astronomer. The Newtonian force "animating" matter and, in the stronger sense of the term, making up the very activity of nature would then be the inheritor of the forces Newton the chemist observed and manipulated, the chemical "affinities" forming and disrupting ever new combinations of matter.[5] The decisive role played by celestial orbits of course remains. Still, at the start of his intense astronomical studies—about 1679—Newton apparently expected to find new forces of attraction only in the heavens, forces *similar* to chemical forces and perhaps easier to study mathematically. Six years later this mathematical study produced an unexpected conclusion: the forces between the planets and those accelerating freely falling bodies are not merely similar but are the same. Attraction is not specific to each planet; it is the same—for the moon circling the earth, for the planets, and even for comets passing through the solar system. Newton set out to discover in the sky forces similar to the chemical forces: the specific affinities, different for each chemical compound and giving each compound qualitatively differentiated activities. What he actually found was a universal law, which, as he emphasized, could be applied to all phenomena—whether chemical, mechanical, or celestial in nature.

The Newtonian synthesis is thus a surprise. It is an unexpected, staggering discovery that the scientific world has commemorated by making Newton the symbol of modern science. What is particularly astonishing is that the basic code of nature appeared to have been cracked in a single creative act.

For a long time this sudden loquaciousness of nature, this triumph of the English Moses, was a source of intellectual scandal for continental rationalists. Newton's work was viewed as a purely empirical discovery that could thus equally well be empirically disproved. In 1747 Euler, Clairaut, and d'Alembert, without doubt some of the greatest scientists of the time, came to the same conclusion: Newton was wrong. In order to describe the moon's motion, a more complex mathematical form must be given to the force of attraction, making it the sum of two terms. For the following two years, each of them believed that nature had proved Newton wrong, and this belief was a source of excitement, not of dismay. Far from considering Newton's discovery synonymous with physical science itself, physicists were blithely contemplating dropping it altogether. D'Alembert went so far as to express scruples about seeking fresh evidence against Newton and giving him *"le coup de pied de l'âne."*[6]

Only one courageous voice against this verdict was raised in France. In 1748, Buffon wrote:

> A physical law is a law only by virtue of the fact that it is easy to measure, and that the scale it represents is not only always the same, but is actually unique. . . . M. Clairaut has raised an objection against Newton's system, but it is at best an objection and must not and cannot become a principle; an attempt should be made to overcome it and not to turn it into a theory the entire consequences of which merely rest on a calculation; for, as I have said, one may represent anything by means of calculation and achieve nothing; and if it is allowed to add one or more terms to a physical law such as that of attraction, we are only adding to arbitrariness instead of representing reality.[7]

Later Buffon was to announce what was to become, although for only a short time, the research program for chemistry:

> The laws of affinity by means of which the constituent parts of different substances separate from others to combine together to form homogeneous substances are the same as the general law governing the reciprocal action of all the celestial bodies on one another: they act in the same way and with the same ratios of mass and distance; a globule of water, of sand or metal acts upon another globule just as the terrestrial globe acts on the moon, and if the laws of affinity have hitherto been regarded as different from those of gravity, it is because they have not been fully understood, not grasped completely; it is because the whole extent of the problem has not been taken in. The figure which, in the case of celestial bodies has little or no effect upon the law of interaction between bodies because of the great distance involved, is, on the contrary, all important when the distance is very small or zero. . . . Our nephews will be able, by calculation, to gain access to this new field of knowledge [that is, to *deduce* the law of interaction between elementary bodies from their figures].[8]

History was to vindicate the naturalist, for whom force was not merely a mathematical artifice but the very essence of the new science of nature. The physicists were later compelled to admit their mistake. Fifty years afterward, Laplace could write his *Système du Monde*. The law of universal gravity had stood all tests successfully: the numerous cases apparently disproving it had been transformed into a brilliant demonstration of its validity. At the same time, under Buffon's influence, the French chemists rediscovered the odd analogy between physical attraction and chemical affinity.[9] Despite the sarcasms of d'Alembert, Condillac, and Condorcet, whose unbending rationalism was quite incompatible with these obscure and barren "analogies," they trod Newton's path in the opposite direction—from the stars to matter.

By the early nineteenth century, the Newtonian program—the reduction of all physicochemical phenomena to the action of forces (in addition to gravitational attraction, this included the repelling force of heat, which makes bodies expand and favors dissolution, as well as electric and magnetic forces)—had become the official program of Laplace's school, which dominated the scientific world at the time when Napoleon dominated Europe.[10]

The early nineteenth century saw the rise of the great French *écoles* and the reorganization of the universities. This is the time when scientists became teachers and professional researchers and took up the task of training their successors.[11] It is also the time of the first attempts to present a synthesis of knowledge, to gather it together in textbooks and works of popularization. Science was no longer discussed in the *salons;* it was taught or popularized.[12] It had become a matter of professional consensus and magistral authority. The first consensus centered around the Newtonian system: in France Buffon's confidence finally triumphed over the rational skepticism of the Enlightenment.

One century after Newton's apotheosis in England, the grandiloquence of these lines written by Ampère's son echoes that of Pope's epitaph:[13]

Announcing the coming of science's Messiah
Kepler had dispelled the clouds around the Arch.
Then the Word was made man, the Word of the seeing God
Whom Plato revered, and He was called Newton.
He came, he revealed the principle supreme,
Eternal, universal, One and unique as God Himself.
The worlds were hushed, he spoke: ATTRACTION.
*This word was the very word of creation.**

For a short time, which nevertheless left an indelible mark, science was triumphant, acknowledged and honored by powerful states and acclaimed as the possessor of a consistent conception of the world. Worshiped by Laplace, Newton became the universal symbol of this golden age. It was a happy moment, indeed, a moment in which scientists were regarded both by themselves and others as the pioneers of progress, achieving an enterprise sustained and fostered by society as a whole.

What is the significance of the Newtonian synthesis today, after the advent of field theory, relativity, and quantum me-

---

*Our translation—authors.

chanics? This is a complex problem, to which we shall return. We now know that nature is not always "conformable and consonant with herself." At the microscopic level, the laws of classical mechanics have been replaced by those of quantum mechanics. Likewise, at the level of the universe, relativistic physics has displaced Newtonian physics. Classical physics nevertheless remains the natural reference point. Moreover, in the sense that we have defined it—that is, as the description of deterministic, reversible, static trajectories—Newtonian dynamics still may be said to form the core of physics.

Of course, since Newton the formulation of classical dynamics has undergone great changes. This was a result of the work of some of the greatest mathematicians and physicists, such as Hamilton and Poincaré. In brief, we may distinguish two periods. First there was a period of clarification and of generalization. During the second period, the very concepts upon which classical dynamics rests, such as initial conditions and the meaning of trajectories, have undergone a critical revision even in the fields in which (in contrast to quantum mechanics and relativity) classical dynamics remains valid. At the moment this book is being written, at the end of the twentieth century, we are still in this second period. Let us turn now to the general language of dynamics that was discovered by nineteenth-century scientists. (In Chapter IX we shall describe briefly the revival of classical dynamics in our time.)

The Language of Dynamics

Today classical dynamics can be formulated in a compact and elegant way. As we shall see, all the properties of a dynamic system can be summarized in terms of a *single* function, the Hamiltonian. The language of dynamics presents a remarkable consistency and completeness. An unambiguous formulation can be given to each "legitimate" problem. No wonder the structure of dynamics has both fascinated and terrified the imagination since the eighteenth century.

In dynamics the same system can be studied from different points of view. In classical dynamics all these points of view are equivalent in the sense that we can go from one to another by a transformation, a change of variables. We may speak of various equivalent representations in which the laws of dynamics are valid. These various equivalent representations form the general language of dynamics. This language can be used to make explicit the static character classical dynamics attributes to the systems it describes: for many classes of dynamic systems, time appears merely as an accident, since their description can be reduced to that of noninteracting mechanical systems. To introduce these concepts in a simple way, let us start with the principle of conservation of energy.

In the ideal world of dynamics, devoid of frictions and collisions, machines have an efficiency of one—the dynamic system comprising the machine merely transmits the whole of the motion it receives. A machine receiving a certain quantity of potential energy (for example, a compressed spring, a raised weight, compressed air) can produce a motion corresponding to an "equal" quantity of kinetic energy, exactly the quantity that would be needed to restore the potential energy the machine has used in producing the motion. The simplest case is that in which the only force considered is gravity (which applies to simple machines, pulleys, levers, capstans, etc.). In this case it is easy to establish an overall relationship of equivalence between cause and effect. The height (h) through which a body falls entirely determines the velocity acquired during its fall. Whether a body of mass m falls vertically, runs down an inclined plane, or follows a roller-coaster path, the acquired velocity (v) and the kinetic energy ($mv^2/2$) depend only on the drop in level h ($v = \sqrt{2gh}$) and enable the body to return to its original height. The work done against the force of gravity implied in this upward motion restores the potential energy, mgh, that the system lost during the fall. Another example is the pendulum, in which kinetic energy and potential energy are periodically transformed into one another.

Of course, if instead of a body falling toward the earth, we are dealing with a system of interacting bodies, the situation is less easily visualized. Still, at each instant the global variation in kinetic energy compensates for the variation in potential energy (bound to the variation in the distances between the points in the system). Here also energy is conserved in an isolated system.

Potential energy (or "potential," conventionally denoted as V), which depends on the relative positions of the particles, is thus a generalization of the quantity that enabled builders of machines to measure the motion a machine could produce as the result of a change in its spatial configuration (for example, the change in the height of a mass m, which is part of the machine, gives it a potential energy mgh). Moreover, potential energy allows us to calculate the set of forces applied at each instant to the different points of the system to be described. At each point the derivative of the potential with respect to the space coordinate q measures the force applied at this point in the direction of that coordinate. Newton's laws of motion thus can be formulated using the potential function instead of force as the main quantity: the variation in the velocity of a point mass at each instant (or the momentum p, the product of the mass and the velocity) is measured by the derivative of the potential with respect to the coordinate q of the mass.

In the nineteenth century this formulation was generalized through the introduction of a new function, the Hamiltonian (H). This function is simply the total energy, the sum of the system's potential and kinetic energy. However, this energy is no longer expressed in terms of positions and velocities, conventionally denoted by q and dq/dt, but in terms of so-called *canonical* variables—coordinates and momenta—for which the standard notation is q and p. In simple cases, such as with a free particle, there is a straightforward relation between velocity and momentum ($p = m\, dq/dt$), but in general the relation is more complicated.

A single function, the Hamiltonian, $H(p, q)$, describes the dynamics of a system *completely.* All our empirical knowledge is put into the form of H. Once this function is known, we may solve, at least in principle, all possible problems. For example, the time variation of the coordinate and of the momenta is simply given by the derivatives of H in respect to p or q. This Hamiltonian formulation of dynamics is one of the greatest achievements in the history of science. It has been progressively extended to cover the theory of electricity and magnetism. It has also been used in quantum mechanics. It is true that in quantum mechanics, as we shall see later, the meaning of the Hamiltonian H had to be generalized: here it is no longer a simple function of the coordinates and momenta, but it becomes a new kind of entity, an operator. (We shall return to this question in Chapter VII.) In any case, the Hamiltonian description is still of the greatest importance today. The equations which, through the derivatives of the Hamiltonian, give the time variation of the coordinates and momenta are the so-called canonical equations. They contain the general properties of all dynamic changes. Here we have the triumph of the mathematization of nature. All dynamic change to which classical dynamics applies can be reduced to these simple mathematical equations.

Using these equations, we can verify the above-mentioned general properties implied by classical dynamics. The canonical equations are *reversible:* time inversion is mathematically the equivalent of velocity inversion. They are also *conservative:* the Hamiltonian, which expressed the system's energy in the canonical variables—coordinates and momenta—is itself conserved by the changes it brings about in the course of time.

We have already noticed that there exist many points of view or "representations" in which the Hamiltonian form of the equations of motion is maintained. They correspond to various choices of coordinates and momenta. One of the basic problems of dynamics is to examine precisely how we can select the pair of canonical variables q and p to obtain as simple a description of dynamics as possible. For example, we could look for canonical variables by which the Hamiltonian is reduced to kinetic energy and depends only on the momenta (and not on the coordinates). What is remarkable is that in this case momenta become constants of motion. Indeed, as we have seen, the time variation of the momenta depends, according to the canonical equation, on the derivative of the Hamiltonian in respect to the coordinates. When this derivative vanishes, the momenta indeed become constants of motion. This is similar to what happens in a "free particle" system. What we have done when we go to a free particle system is "eliminate" the interaction through a change of representation. We will define systems for which this is possible as "integrable systems." Any integrable system may thus be represented as a set of units, each changing in isolation, quite independently of all the others, in that eternal and immutable motion Aristotle attributed to the heavenly bodies (Figure 1).

We have already noted that in dynamics "everything is given." Here this means that, from the very first instant, the value of the various invariants of the motion is fixed; nothing

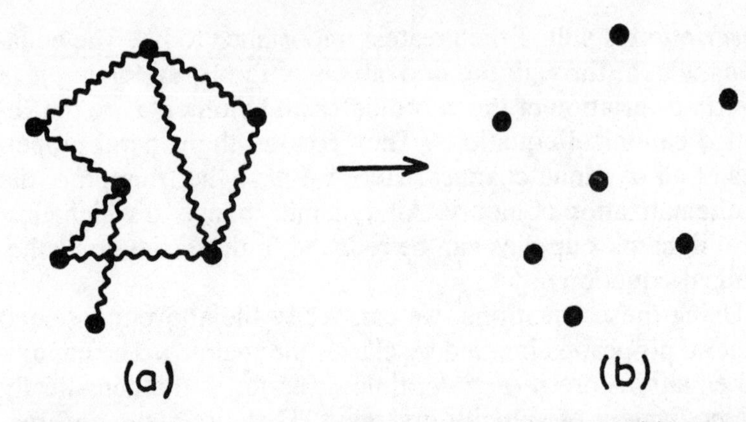

Figure 1. Two representations of the same dynamic system: (a) as a set of interacting points; the interaction between the points is represented by wavy lines; (b) as a set where each point behaves independently from the others. The potential energy being eliminated, their respective motions are not explicitly dependent on their relative positions.

may "happen" or "take place." Here we reach one of those dramatic moments in the history of science when the description of nature was nearly reduced to a static picture. Indeed, through a clever change of variables, all interaction could be made to disappear. It was believed that integrable systems, reducible to free particles, were the prototype of dynamic systems. Generations of physicists and mathematicians tried hard to find for each kind of systems the "right" variables that would eliminate the interactions. One widely studied example was the three-body problem, perhaps the most important problem in the history of dynamics. The moon's motion, influenced by both the earth and the sun, is one instance of this problem. Countless attempts were made to express it in the form of an integrable system until, at the end of the nineteenth century, Bruns and Poincaré showed that this was impossible. This came as a surprise and, in fact, announced the end of all simple extrapolations of dynamics based on integrable systems. The discovery of Bruns and Poincaré shows that dynamic systems are not isomorphic. Simple, integrable systems can indeed be reduced to noninteracting units, but in general, interactions cannot be eliminated. Although this discovery was not clearly understood at the time, it implied the demise of the conviction that the dynamic world is homogeneous, reducible to the concept of integrable systems. Nature as an evolving, interactive multiplicity thus resisted its reduction to a timeless and universal scheme.

There were other indications pointing in the same direction. We have mentioned that trajectories correspond to deterministic laws; once an initial state is given, the dynamic laws of motion permit the calculation of trajectories at each point in the future or the past. However, a trajectory may become intrinsically indeterminate at certain singular points. For instance, a rigid pendulum may display two qualitatively different types of behavior—it may either oscillate or swing around its points of suspension. If the initial push is just enough to bring it into a vertical position with zero velocity, the direction in which it will fall, and therefore the nature of its motion, are indeterminate. An infinitesimal perturbation would be enough to set it rotating or oscillating. (This problem of the "instability" of motion will be discussed fully in Chapter IX.)

It is significant that Maxwell had already stressed the importance of these singular points. After describing the explosion of gun cotton, he goes on to say:

> In all such cases there is one common circumstance—the system has a quantity of potential energy, which is capable of being transformed into motion, but which cannot begin to be so transformed till the system has reached a certain configuration, to attain which requires an expenditure of work, which in certain cases may be infinitesimally small, and in general bears no definite proportion to the energy developed in consequence thereof. For example, the rock loosed by frost and balanced on a singular point of the mountain-side, the little spark which kindles the great forest, the little word which sets the world a fighting, the little scruple which prevents a man from doing his will, the little spore which blights all the potatoes, the little gemmule which makes us philosophers or idiots. Every existence above a certain rank has its singular points: the higher the rank, the more of them. At these points, influences whose physical magnitude is too small to be taken account of by a finite being, may produce results of the greatest importance. All great results produced by human endeavour depend on taking advantage of these singular states when they occur.[14]

This conception received no further elaboration owing to the absence of suitable mathematical techniques for identifying systems containing such singular points and the absence of the chemical and biological knowledge that today affords, as we shall see later, a deeper insight into the truly essential role played by such singular points.

Be that as it may, from the time of Leibniz' monads (see the conclusion to section 4) down to the present day (for example, the stationary states of the electrons in the Bohr model—see Chapter VII), integrable systems have been the model par excellence of dynamic systems, and physicists have attempted to extend the properties of what is actually a very special class of Hamiltonian equations to cover all natural processes. This is quite understandable. The class of integrable systems is the only one that, until recently, had been thoroughly explored. Moreover, there is the fascination always associated with a closed system capable of posing all problems, provided it does not define them as meaningless. Dynamics is such a language; being complete, it is by definition coextensive with the world it is describing. It assumes that all problems, whether simple or complex, resemble one another since it can always pose them in the same general form. Thus the temptation to conclude that all problems resemble one another from the point of view of their solutions as well, and that nothing new can appear as a result of the greater or lesser complexity of the integration procedure. It is this intrinsic homogeneity that we now know to be false. Moreover, the mechanical world view was acceptable as long as all observables referred in one way or another to motion. This is no longer the case. For example, unstable particles have an energy that can be related to motion but that also has a lifetime that is a quite different type of observable, more closely related to irreversible processes, as we shall describe them in Chapters IV and V. The necessity of introducing new observables into the theoretical sciences was, and still is today, one of the driving forces that move us beyond the mechanical world view.

Laplace's Demon

Extrapolations from the dynamic description discussed above have a symbol—the demon imagined by Laplace, capable at any given instant of observing the position and velocity of each mass that forms part of the universe and of inferring its evolution, both toward the past and toward the future. Of course, no one has ever dreamed that a physicist might one day benefit from the knowledge possessed by Laplace's demon. Laplace himself only used this fiction to demonstrate the extent of our ignorance and the need for a statistical description of certain processes. The problematics of Laplace's demon are not related to the question of whether a deterministic prediction of the course of events is actually possible, but whether it is possible in principle, *de jure*. This possibility seems to be implied in mechanistic description, with its characteristic duality based on dynamic law and initial conditions.

Indeed, the fact that a dynamic system is governed by a deterministic *law*, even though in practice our ignorance of the initial state precludes any possibility of deterministic predictions, allows the "objective truth" of the system as it would be seen by Laplace's demon to be distinguished from empirical limitations due to our ignorance. In the context of classical dynamics, a deterministic description may be unattainable in practice; nevertheless, it stands as a *limit* that defines a series of increasingly accurate descriptions.

It is precisely the consistency of this duality formed by dynamic law and initial conditions that is challenged in the revival of classical mechanics, which we will describe in Chapter IX. We shall see that the motion may become so complex, the trajectories so varied, that no observation, whatever its precision, can lead us to the determination of the exact initial conditions. But at that point the duality on which classical mechanics was constructed breaks down. We can predict only the average behavior of bundles of trajectories.

Modern science was born out of the breakdown of the animistic alliance with nature. Man seemed to possess a place in the Aristotelian world as both a living and a knowing creature.

The world was made to his measure. The first experimental dialogue received part of its social and philosophic justification from another alliance, this time with the rational God of Christianity. To the extent to which dynamics has become and still is the model of science, certain implications of this historical situation have persisted to our day.

Science is still the prophetic announcement of a description of the world seen from a divine or demonic point of view. It is the science of Newton, the new Moses to whom the truth of the world was unveiled; it is a *revealed* science that seems alien to any social and historical context identifying it as the result of the activity of human society. This type of inspired discourse is found throughout the history of physics. It has accompanied each conceptual innovation, each occasion at which physics seemed at the point of unification and the prudent mask of positivism was dropped. Each time physicists repeated what Ampère's son stated so explicitly: this word—universal attraction, energy, field theory, or elementary particles—is the word of creation. Each time—in Laplace's time, at the end of the nineteenth century, or even today—physicists announced that physics was a closed book or about to become so. There was only one final stronghold where nature continued to resist, the fall of which would leave it defenseless, conquered, and subdued by our knowledge. They were thus unwittingly repeating the ritual of the ancient faith. They were announcing the coming of the new Moses, and with him a new Messianic period in science.

Some might wish to disregard this prophetic claim, this somewhat naïve enthusiasm, and it is certainly true that dialogue with nature has gone on all the same, together with a search for new theoretical languages, new questions, and new answers. But we do not accept a rigid separation between the scientist's "actual" work and the way he judges, interprets, and orientates this work. To accept it would be to reduce science to an ahistorical accumulation of results and to pay no attention to what scientists are looking for, the ideal knowledge they try to attain, the reasons why they occasionally quarrel or remain unable to communicate with each other.[15]

Once again, it was Einstein who formulated the enigma produced by the myth of modern science. He has stated that the miracle, the only truly astonishing feature, is that science exists at all, that we find a convergence between nature and the human mind. Similarly, when, at the end of the nineteenth century, du Bois Reymond made Laplace's demon the very incarnation of the logic of modern science, he added, "Ignoramus, ignorabimus": we shall always be totally ignorant of the relationship between the world of science and the mind which knows, perceives, and creates this science.[16]

Nature speaks with a thousand voices, and we have only begun to listen. Nevertheless, for nearly two centuries Laplace's demon has plagued our imagination, bringing a nightmare in which all things are insignificant. If it were really true that the world is such that a demon—a being that is, after all, like us, possessing the same science, but endowed with sharper senses and greater powers of calculation—could, starting from the observation of an instantaneous state, calculate its future and past, if nothing qualitatively differentiates the simple systems we can describe from the more complex ones for which a demon is needed, then the world is nothing but an immense tautology. This is the challenge of the science we have inherited from our predecessors, the spell we have to exorcise today.

CHAPTER III
THE TWO CULTURES

Diderot and the Discourse of the Living

In his interesting book on the history of the idea of progress, Nisbet writes:

> No single idea has been more important than, perhaps as important as, the idea of progress in Western civilization for nearly three thousand years.[1]

There has been no stronger support for the idea of progress than the accumulation of knowledge. The grandiose spectacle of this gradual increase of knowledge is indeed a magnificent example of a successful collective human endeavor.

Let us recall the remarkable discoveries achieved at the end of the eighteenth century and the beginning of the nineteenth century: the theories of heat, electricity, magnetism, and optics. It is not surprising that the idea of scientific progress, already clearly formulated in the eighteenth century, dominated the nineteenth. Still, as we have pointed out, the position of science in Western culture remained unstable. This lends a dramatic aspect to the history of ideas from the high point of the Enlightenment.

We have already stated the alternative: to accept science with what appears to be its alienating conclusions or to turn to an antiscientific metaphysics. We have also emphasized the solitude felt by modern men, the loneliness described by Pascal, Kierkegaard, or Monod. We have mentioned the antiscientific implications of Heidegger's metaphysics. Now we wish to discuss more fully some aspects of the intellectual history of the West, from Diderot, Kant, and Hegel to Whitehead and Bergson; all of them attempted to analyze and limit the scope of modern science as well as to open new perspectives seen as radically alien to that science. Today it is usually agreed that those attempts have for the most part failed. Few would accept, for example, Kant's division of the world into phenomenal and noumenal spheres, or Bergson's "intuition" as an alternative path to a knowledge whose significance would parallel that of science. Still these attempts are part of our heritage. The history of ideas cannot be understood without reference to them.

We shall also briefly discuss scientific positivism, which is based on the separation of what is true from what is scientifically useful. At the outset this positivistic view may seem to oppose clearly the metaphysical views we have mentioned, views that I. Berlin described as the "Counter-Enlightenment." However, their fundamental conclusion is the same: we must reject science as a basis for true knowledge even if at the same time we recognize its practical importance or we deny, as positivists do, the possibility of any other cognitive enterprise.

We must remember all these developments to understand what is at stake. To what extent is science a basis for the intelligibility of nature, including man? What is the meaning of the idea of progress today?

Diderot, one of the towering figures of the Enlightenment, is certainly no representative of antiscientific thought. On the contrary, his confidence in science, in the possibilities of knowledge, was total. Yet this is the very reason why science had, following Diderot, to understand life before it could hope to achieve any coherent vision of nature.

We have already mentioned that the birth of modern science was marked by the abandonment of vitalist inspiration and, in particular, of Aristotelian final causes. However, the issue of the organization of living matter remained and became a challenge for classical science. Diderot, at the height of the Newtonian triumph, emphasizes that this problem was repressed by physics. He imagines it as haunting the dreams of physicists who cannot conceive of it while they are awake. The physicist d'Alembert is dreaming:

> "A living point . . . No, that's wrong. Nothing at all to begin with, and then a living point. This living point is joined by another, and then another, and from these successive joinings there results a unified being, for I am a

unity, of that I am certain. . . . (As he said this he felt himself all over.) But how did this unity come about?"

"Now listen, Mr. Philosopher, I can understand an aggregate, a tissue of tiny sensitive bodies, but an animal! . . . A whole, a system that is a unit, an individual conscious of its own unity! I can't see it, no, I can't see it."[2]

In an imaginary conversation with d'Alembert, Diderot speaks in the first person, demonstrating the inadequacy of mechanistic explanation:

> Look at this egg: with it you can overthrow all the schools of theology and all the churches in the world. What is this egg? An insensitive mass before the germ is put into it . . . How does this mass evolve into a new organization, into sensitivity, into life? Through heat. What will generate heat in it? Motion. What will the successive effects of motion be? Instead of answering me, sit down and let us follow out these effects with our eyes from one moment to the next. First there is a speck which moves about, a thread growing and taking colour, flesh being formed, a beak, wing-tips, eyes, feet coming into view, a yellowish substance which unwinds and turns into intestines—and you have a living creature. . . . Now the wall is breached and the bird emerges, walks, flies, feels pain, runs away, comes back again, complains, suffers, loves, desires, enjoys, it experiences all your affections and does all the things you do. And will you maintain, with Descartes, that it is an imitating machine pure and simple? Why, even little children will laugh at you, and philosophers will answer that if it is a machine you are one too! If, however, you admit that the only difference between you and an animal is one of organization, you will be showing sense and reason and be acting in good faith; but then it will be concluded, contrary to what you had said, that from an inert substance arranged in a certain way and impregnated by another inert substance, subjected to heat and motion, you will get sensitivity, life, memory, consciousness, passions, thought . . . Just listen to your own arguments and you will feel how pitiful

> they are. You will come to feel that by refusing to entertain a simple hypothesis that explains everything—sensitivity as a property common to all matter or as a result of the organization of matter—you are flying in the face of common sense and plunging into a chasm of mysteries, contradictions and absurdities.[3]

In opposition to rational mechanics, to the claim that material nature is nothing but inert mass and motion, Diderot appeals to one of physics' most ancient sources of inspiration, namely, the growth, differentiation, and organization of the embryo. Flesh forms, and so does the beak, the eyes, and the intestines; a gradual organization occurs in biological "space," out of an apparently homogeneous environment differentiated forms appear at exactly the right time and place through the effects of complex and coordinated processes.

How can an inert mass, even a Newtonian mass animated by the forces of gravitational interaction, be the starting point for organized active local structures? We have seen that the Newtonian system is a world system: no local configuration of bodies can claim a particular identity; none is more than a contingent proximity between bodies connected by general relations.

But Diderot does not despair. Science is only beginning; rational mechanics is merely a first, overly abstract attempt. The spectacle of the embryo is enough to refute its claims to universality. This is why Diderot compares the work of great "mathematicians" such as Euler, Bernoulli, and d'Alembert to the pyramids of the Egyptians, awe-inspiring witnesses to the genius of their builders, now lifeless ruins, alone and forlorn. True science, alive and fruitful, will be carried on elsewhere.[4]

Moreover, it seems to him that this new science of organized living matter has already begun. His friend d'Holbach is busy studying chemistry, Diderot himself has chosen medicine. The problem in chemistry as well as in medicine is to replace inert matter with *active* matter capable of organizing itself and producing living beings. Diderot claims that matter has to be sensitive. Even a stone has sensation in the sense that the molecules of which it is composed actively seek certain combinations rather than others and thus are governed their likes and dislikes. The sensitivity of the whole organism is then simply the sum of that of its parts, just as a swarm of bees with its globally coherent behavior is the result of interactions between one bee and another; and, Diderot thereby concludes, the human soul does not exist any more than the soul of the beehive does.[5]

Diderot's vitalist protest against physics and the universal laws of motion thus stems from his rejection of any form of spiritualist dualism. Nature must be described in such a way that man's very existence becomes understandable. Otherwise, and this is what happens in the mechanistic world view, the scientific description of nature will have its counterpart in man as an automaton endowed with a soul and thereby alien to nature.

The twofold basis of materialistic naturalism, at once chemical and medical, that Diderot employed to counter the physics of his time is recurrent in the eighteenth century. While biologists speculated about the animal-machine, the preexistence of germs, and the chain of living creatures—all problems close to theology[6]—chemists and physicians had to face directly the complexity of real processes in both chemistry and life. Chemistry and medicine were, in the late eighteenth century, privileged sciences for those who fought against the physicists' *esprit de système* in favor of a science that would take into account the diversity of natural processes. A physicist could be pure *esprit,* a precocious child, but a physician or a chemist must be a man of experience: he must be able to decipher the signs, to spot the clues. In this sense, chemistry and medicine are arts. They demand judgment, application, and tenacious observation. Chemistry is a madman's passion, Venel concluded in the article he wrote for Diderot's *Encyclopédie,* an eloquent defense of chemistry against the abstract imperialism of the Newtonians.[7] To emphasize the fact that protests raised by chemists and physicians against the way physicists reduced living processes to peaceful mechanisms and the quiet unfolding of universal laws were common in Diderot's day, we invoke the eminent figure of Stahl, the father of vitalism and inventor of the first consistent chemical systematics.

According to Stahl, universal laws apply to the living only in the sense that these laws condemn them to death and corruption; the matter of which living beings are composed is so frail, so easily decomposed, that if it were governed solely by the common laws of matter, it would not withstand decay or dissolution for a moment. If a living creature is to survive in spite of the general laws of physics, however short its life when it is compared to that of a stone or another inanimate object, it has to possess in itself a "principle of conservation" that maintains the harmonious equilibrium of the texture and structure of its body. The astonishing longevity of a living body in view of the extreme corruptibility of its constitutive matter is thus indicative of the action of a "natural, permanent, immanent principle," of a particular cause that is alien to the laws of inanimate matter and that constantly struggles against the constantly active corruption whose inevitability these laws imply.[8]

To us this analysis of life sounds both near and remote. It is close to us in its acute awareness of the singularity and the precariousness of life. It is remote because, like Aristotle, Stahl defined life in static terms, in terms of conservation, not of becoming or evolution. Still, the terminology used by Stahl can be found in recent biological literature, for example, where we read that enzymes "combat" decay and allow the body to ward off the death to which it is inexorably doomed by physics. Here also, biological organization defies the laws of nature, and the only "normal" trend is that which leads to death (see Chapter V).

Indeed, Stahl's vitalism is relevant as long as the laws of physics are identified with evolution toward decay and disorganization. Today the "vitalist principle" has been superseded by the succession of improbable mutations preserved in the genetic message "governing" the living structure. Nonetheless, some extrapolations starting from molecular biology relegate life to the confines of nature—that is, conclude life is compatible with the basic laws of physics but purely contingent. This was explicitly stated by Monod: life does not "follow from the laws of physics, it is compatible with them. Life is an event whose singularity we have to recognize."

But the transition from matter to life can also be viewed in a different way. As we shall see, far from equilibrium, new self-organizational processes arise. (These questions will be studied in detail in Chapters V and VI.) In this way biological organization begins to appear as a *natural* process.

However, long before these recent developments, the problematics of life had been transformed. In a politically transformed Europe the intellectual landscape was remodeled as the Romantic movement, closely linked with the Counter-Enlightenment, shows.

Stahl criticized the metaphor of the automaton because, unlike a living being, the purpose of an automaton does not lie within itself; its organization is imposed upon it by its maker. Diderot, far from situating the study of life outside the reach of science, saw it as representing the future of a science he considered to be still in its infancy. A few years later, such points of view were to be challenged.[9] Mechanical change, activity as described by the laws of motion, had now become synonymous with the artificial and with death. Opposed to it, united in a complex with which we are now quite familiar, were the concepts of life, spontaneity, freedom, and spirit. This opposition was paralleled by the opposition between calculation and manipulation on the one hand, and the free speculative activity of the mind on the other. Through speculation the philosopher would reach the spiritual activity at the core of nature. As for the scientist, his concern with nature would be reduced to taking it as a set of manipulable and measurable objects; he would thus be able to take possession of nature, to dominate and control it but not understand it. Thus the intelligibility of nature would lie beyond the grasp of science.

We are not concerned here with the history of philosophy but merely with emphasizing the extent to which the philosophical criticism of science had at this time become harsher, resembling certain modern forms of antiscience. It was no longer a question of refuting rather naïve and shortsighted generalizations that only have to be repeated aloud—to use Diderot's language—to make even children laugh, but of refuting the type of approach that produced experimental and mathematical knowledge of nature. Scientific knowledge is not being criticized for its limitations but for its nature, and a *rival* knowledge, based on another approach, is being announced. Knowledge is fragmented into two opposed modes of inquiry.

From a philosophical point of view, the transition from Diderot to the Romantics and, more precisely, from one of these two types of critical attitudes toward science to the other, can be found in Kant's transcendental philosophy, the essential point being that the Kantian critique identified science in general with its Newtonian realization. It thereby branded as impossible any opposition to classical science that was not an opposition to science itself. Any criticism against Newtonian physics must then be seen as aimed at downgrading the rational understanding of nature in favor of a different form of knowledge. Kant's approach had immense repercussions, which continue down to our day. Let us therefore summarize his point of view as presented in *Critique of Pure Reason,* which, in opposition to the progressist views of the Enlightenment, presents the closed and limiting conception of science we have just defined.

Kant's Critical Ratification

How to restore order in the intellectual landscape left in disarray with the disappearance of God conceived as the rational principle that links science and nature? How could scientists ever have access to global truth when it could no longer be asserted, except metaphorically, that science deciphers the word of creation? God was now silent or at least no longer spoke the same language as human reason. Moreover, in a nature from which time was eliminated, what remained of our subjective experience? What was the meaning of freedom, destiny, or ethical values?

Kant argued that there were two levels of reality: a phenomenal level that corresponds to science, and a noumenal level corresponding to ethics. The phenomenal order is created by the human mind. The noumenal level transcends man's intellect; it corresponds to a spiritual reality that supports his ethical and religious life. In a way, Kant's solution is the only one possible for those who assert both the reality of ethics and the reality of the objective world as it is expressed by classical science. Instead of God, it is now man himself who is the source of the order he perceives in nature. Kant justifies both scientific knowledge and man's alienation from the phenomenal world described by science. From this perspective we can see that Kantian philosophy explicitly spells out the philosophical content of classical science.

Kant defines the subject of critical philosophy as *transcendental*. It is not concerned with the objects of experience but is based on the a priori fact that a systematic knowledge of these objects is possible (this is for him proved by the existence of physics), going on to state the a priori conditions of possibility for this mode of knowledge.

To do so a distinction must be made between the direct sensations we receive from the outside world and the objective, "rational" mode of knowledge. Objective knowledge is not passive; it forms its objects. When we take a phenomenon as the object of experience, we assume a priori before we actually experience it that it obeys a given set of principles. Insofar as it is perceived as a possible object of knowledge, it is the product of our mind's synthetic activity. We find ourselves in the objects of our knowledge, and the scientist himself is thus the source of the universal laws he discovers in nature.

The a priori conditions of experience are also the conditions for the existence of the objects of experience. This celebrated statement sums up the "Copernican revolution" achieved by Kant's "transcendental" inquiry. The subject no longer "revolves" around its object, seeking to discover the laws by which it is governed or the language by which it may be deciphered. Now the subject itself is at the center, imposing its laws, and the world perceived speaks the language of that subject. No wonder, then, that Newtonian science is able to describe the world from an external, almost divine point of view!

That all perceived phenomena are governed by the laws of our mind does not mean that a concrete knowledge of these objects is useless. According to Kant, science does not engage in a dialogue with nature but imposes its own language upon it. Still it must discover, in each case, the specific message expressed in this general language. A knowledge of the a priori concepts alone is vain and empty.

From the Kantian point of view Laplace's demon, the symbol of the scientific myth, is an illusion, but it is a *rational* illusion. Although it is the result of a limiting process and, as such, illegitimate, it is still the expression of a legitimate conviction that is the driving force of science—the conviction that, in its entirety, nature is rightfully subjected to the laws that scientists succeed in deciphering. Wherever it goes, whatever it questions, science will always obtain, if not the same answer, at least the same kind of answer. There exists a single universal syntax that includes all possible answers.

Transcendental philosophy thus ratified the physicist's

claim to have found the definitive form of all positive knowledge. At the same time, however, it secured for philosophy a dominant position in respect to science. It was no longer necessary to look for the philosophic significance of the results of scientific activity. From the transcendental standpoint, those results cannot lead to anything really new. It is science, not its results, that is the subject of philosophy; science taken as a repetitive and closed enterprise provides a stable foundation for transcendental reflection.

Therefore, while it ratifies all the claims of science, Kant's critical philosophy actually limits scientific activity to problems that can be considered both easy and futile. It condemns science to the tedious task of deciphering the monotonous language of phenomena while keeping for itself questions of human "destiny": what man may know, what he must do, what he may hope for. The world studied by science, the world accessible to positive knowledge is "only" the world of phenomena. Not only is the scientist unable to know things in themselves, but even the questions he asks are irrelevant to the real problems of mankind. Beauty, freedom, and ethics cannot be objects of positive knowledge. They belong to the noumenal world, which is the domain of philosophy, and they are quite unrelated to the phenomenal world.

We can accept Kant's starting point, his emphasis on the active role man plays in scientific description. Much has already been said about experimentation as the art of choosing situations that are hypothetically governed by the law under investigation and staging them to give clear, experimental answers. For each experiment certain principles are presupposed and thus cannot be established by that experiment. However, as we have seen, Kant goes much further. He denies the diversity of possible scientific points of view, the diversity of presupposed principles. In agreement with the myth of classical science, Kant is after the *unique* language that science deciphers in nature, the unique set of a priori principles on which physics is based and that are thus to be identified with the categories of human understanding. Thus Kant denies the need for the scientist's active choice, the need for a selection of a problematic situation corresponding to a particular theoretical language in which definite questions may be asked and experimental answers sought.

Kant's critical ratification defines scientific endeavor as silent and systematic, closed within itself. By so doing, philosophy endorses and perpetuates the rift, debasing and surrendering the whole field of positive knowledge to science while retaining for itself the field of freedom and ethics, conceived as alien to nature.

A Philosophy of Nature? Hegel and Bergson

The Kantian truce between science and philosophy was a fragile one. Post-Kantian philosophers disrupted this truce in favor of a new philosophy of science, presupposing a new path to knowledge that was distinct from science and actually hostile to it. Speculation released from the constraints of any experimental dialogue reigned supreme, with disastrous consequences for the dialogue between scientists and philosophers. For most scientists, the philosophy of science became synonymous with arrogant, absurd speculation riding roughshod over facts, and indeed regularly proven wrong by the facts. On the other side, for most philosophers it has become a symbol of the dangers involved in dealing with nature and in competing with science. The rift among science, philosophy, and humanistic studies was thus made greater by mutual disdain and fear.

As an example of this speculative approach to nature, let us first consider Hegel. Hegel's philosophy has cosmic dimensions. In his system increasing levels of complexity are specified, and nature's purpose is the eventual self-realization of its spiritual element. Nature's history is fulfilled with the appearance of man—that is, with the coming of Spirit apprehending itself.

The Hegelian philosophy of nature systematically incorporates all that is denied by Newtonian science. In particular, it rests on the qualitative difference between the simple behavior described by mechanics and the behavior of more complex entities such as living beings. It denies the possibility of reducing those levels, rejecting the idea that differences are merely apparent and that nature is basically homogeneous and simple. It affirms the existence of a hierarchy, each level of which presupposes the preceding ones.

Unlike the Newtonian authors of *romans de la matière,* of world-embracing panoramas ranging from gravitational interactions to human passions, Hegel knew perfectly well that his distinctions among levels (which, quite apart from his own interpretation, we may acknowledge as corresponding to the idea of an increasing complexity in nature and to a concept of time whose significance would be richer on each new level) ran *counter* to his day's mathematical science of nature. He therefore set out to limit the significance of this science, to show that mathematical description is restricted to the most trivial situations. Mechanics can be mathematized because it attributes only space-time properties to matter. "A brick does not kill a man merely because it is a brick, but solely because of its acquired velocity; this means that the man is killed by space and time."[10] The man is killed by what we call kinetic energy ($mv^2/2$)—by an abstract quantity defining mass and velocity as interchangeable; the same murderous effect can be achieved by reducing one and increasing the other.

It is precisely this interchangeability that Hegel sets as a condition for mathematization that is no longer satisfied when the mechanical level of description is abandoned for a "higher" one involving a larger spectrum of physical properties.

In a sense Hegel's system provides a consistent philosophic response to the crucial problems of time and complexity. However, for generations of scientists it represented the epitome of abhorrence and contempt. In a few years, the intrinsic difficulties of Hegel's philosophy of nature were aggravated by the obsolescence of the scientific background on which his system was based, for Hegel, of course, based his rejection of the Newtonian system on the scientific conceptions of his time.[11] And it was precisely those conceptions that were to fall into oblivion with astonishing speed. It is difficult to imagine a less opportune time than the beginning of the nineteenth century for seeking experimental and theoretical support for an alternative to classical science. Although this time was characterized by a remarkable extension of the experimental scope of science (see Chapter IV) and by a proliferation of theories that seemed to contradict Newtonian science, most of those theories had to be given up only a few years after their appearance.

At the end of the nineteenth century, when Bergson undertook his search for an acceptable alternative to the science of his time, he turned to intuition as a form of speculative knowledge, but he presented it as quite different from that of the Romantics. He explicitly stated that intuition is unable to produce a system but produces only results that are always partial and nongeneralizable, results to be formulated with great caution. In contrast, generalization is an attribute of "intelligence," the greatest achievement of which is classical science. Bergsonian intuition is a concentrated attention, an increasingly difficult attempt to penetrate deeper into the singularity of things. Of course, to communicate, intuition must have recourse to language—"in order to be transmitted, it will have to use ideas as a conveyance."[12] This it does with infinite patience and circumspection, at the same time accumulating images and comparisons in order to "embrace reality,"[13] thus suggesting in an increasingly precise way what cannot be communicated by means of general terms and abstract ideas.

Science and intuitive metaphysics "are or can become equally precise and definite. They both bear upon reality itself. But each one of them retains only half of it so that one could see in them, if one wished, two subdivisions of science or two departments of metaphysics, if they did not mark divergent directions of the activity of thought."[14]

The definition of these two divergent directions may also be considered as the historical consequence of scientific evolution. For Bergson, it is no longer a question of finding scientific alternatives to the physics of his time. In his view, chemistry and biology had definitely chosen mechanics as their model. The hopes that Diderot had cherished for the future of chemistry and medicine had thus been dashed. In Bergson's view, science is a whole and must therefore be judged as a whole. And this is what he does when he presents science as the product of a practical intelligence whose aim is to dominate matter and that develops by abstraction and generalization the intellectual categories needed to achieve this domination. Science is the product of our vital need to exploit the world, and its concepts are determined by the necessity of manipulating objects, of making predictions, and of achieving reproducible actions. This is why rational mechanics represents the very essence of science, its actual embodiment. The other sciences are more vague, awkward manifestations of an approach that is all the more successful the more inert and disorganized the terrain it explores.

For Bergson all the limitations of scientific rationality can be reduced to a single and decisive one: it is incapable of understanding *duration* since it reduces time to a sequence of instantaneous states linked by a deterministic law.

"Time is invention, or it is nothing at all."[15] Nature is change, the continual elaboration of the new, a totality being created in an essentially open process of development without any preestablished model. "Life progresses and endures in time."[16] The only part of this progression that intelligence can grasp is what it succeeds in fixing in the form of manipulable and calculable elements and in referring to a time seen as sheer juxtaposition of instants.

> Therefore, physics "is limited to coupling simultaneities between the events that make up this time and the positions of the mobile *T* on its trajectory. It detaches these events from the whole, which at every moment puts on a new form and which communicates to them something of its novelty. It considers them in the abstract, such as they would be outside of the living whole, that is to say, in a time unrolled in space. It retains only the events or systems of events that can be thus isolated without being made to undergo too profound a deformation, because only these lend themselves to the application of its method. Our physics dates from the day when it was known how to isolate such systems."[17]

When it comes to understanding duration itself, science is powerless. What is needed is intuition, a "direct vision of the mind by the mind."[18] "Pure change, real duration, is something spiritual. Intuition is what attains the spirit, duration, pure change.[19]

Can we say Bergson has failed in the same way that the post-Kantian philosophy of nature failed? He has failed insofar as the metaphysics based on intuition he wished to create has not materialized. He has not failed in that, unlike Hegel, he had the good fortune to pass judgment upon science that was, on the whole, firmly established—that is, classical science at its apotheosis, and thus identified problems which are indeed still our problems. But, like the post-Kantian critics, he identified the science of his time with science in general. He thus attributed to science *de jure* limitations that were only *de facto*. As a consequence he tried to define once and for all a *statu quo* for the respective domains of science and other intellectual activities. Thus the only perspective remaining open for him was to introduce a way in which antagonistic approaches could at best merely coexist.

In conclusion, even if the way in which Bergson sums up the achievement of classical science is still to some extent acceptable, we can no longer accept it as a statement of the eternal limits of the scientific enterprise. We conceive of it more as a program that is beginning to be implemented by the metamorphosis science is now undergoing. In particular, we know that time linked with motion does not exhaust the meaning of time in physics. Thus the limitations Bergson criticized are beginning to be overcome, not by abandoning the scientific approach or abstract thinking but by perceiving the limitations of the concepts of classical dynamics and by discovering new formulations valid in more general situations.

Process and Reality: Whitehead

As we have emphasized, the element common to Kant, Hegel, and Bergson is the search for an approach to reality that is different from the approach of classical science. This is also the fundamental aim of Whitehead's philosophy, which is resolutely pre-Kantian. In his most important book, *Process and Reality,* he puts us back in touch with the great philosophies of the Classical Age and their quest for rigorous conceptual experimentation.

Whitehead sought to understand human experience as a process belonging to nature, as physical existence. This challenge led him, on the one hand, to reject the philosophic tradition that defined subjective experience in terms of consciousness, thought, and sense perception, and, on the other, to conceive of all *physical* existence in terms of enjoyment, feeling, urge, appetite, and yearning—that is, to cross swords with what he calls "scientific materialism," born in the seventeenth century. Like Bergson, Whitehead was thus led to point out the basic inadequacies of the theoretical scheme developed by seventeenth-century science:

> The seventeenth century had finally produced a scheme of scientific thought framed by mathematicians, for the use of mathematicians. The great characteristic of the mathematical mind is its capacity for dealing with abstractions; and for eliciting from them clear-cut demonstrative trains of reasoning, entirely satisfactory so long as it is those abstractions which you want to think about. The enormous success of the scientific abstractions, yielding on the one hand matter with its simple location in space and time, on the other hand mind, perceiving, suffering, reasoning, but not interfering, has foisted on to philosophy the task of accepting them as the most concrete rendering of fact.
>
> Thereby, modern philosophy has been ruined. It has oscillated in a complex manner between three extremes. There are the dualists, who accept matter and mind as on equal basis, and the two varieties of monists, those who put mind inside matter, and those who put matter inside mind. But this juggling with abstractions can never overcome the inherent confusion introduced by the ascription of misplaced concreteness to the scientific scheme of the seventeenth century.[20]

However, Whitehead considered this to be only a temporary situation. Science is not doomed to remain a prisoner of confusion.

We have already raised the question of whether it is possible to formulate a philosophy of nature that is not directed against science. Whitehead's cosmology is the most ambitious attempt to do so. Whitehead saw no basic contradiction between science and philosophy. His purpose was to define the conceptual field within which the problem of human experience and physical processes could be dealt with consistently and to determine the conditions under which the problem could be solved. What had to be done was to formulate the principles necessary to characterize all forms of existence, from that of stones to that of man. It is precisely this universality that, in Whitehead's opinion, defines his enterprise as "philosophy." While each scientific theory selects and abstracts from the world's complexity a peculiar set of relations, philosophy cannot favor any particular region of human experience. Through conceptual experimentation it must construct a consistency that can accommodate all dimensions of experience, whether they belong to physics, physiology, psychology, biology, ethics, etc.

Whitehead understood perhaps more sharply than anyone else that the creative evolution of nature could never be conceived if the elements composing it were defined as permanent, individual entities that maintained their identity throughout all changes and interactions. But he also understood that to make all permanence illusory, to deny being in the name of becoming, to reject entities in favor of a continuous and ever-changing flux meant falling once again into the trap always lying in wait for philosophy—to "indulge in brilliant feats of explaining away."[21]

Thus for Whitehead the task of philosophy was to reconcile permanence and change, to conceive of things as processes, to demonstrate that becoming forms entities, individual identities that are born and die. It is beyond the scope of this book to give a detailed presentation of Whitehead's system. Let us only emphasize that he demonstrated the connection between a philosophy of *relation*—no element of nature is a permanent support for changing relations; each receives its identity from its relations with others—and a philosophy of *innovating becoming*. In the process of its genesis, each existent unifies the multiplicity of the world, since it adds to this multiplicity an extra set of relations. At the creation of each new entity "the many become one and are increased by one."[22]

In the conclusion of this book, we shall again encounter Whitehead's question of permanence and change, this time as it is raised in physics; we shall speak of entities formed by their irreversible interaction with the world. Today physics has discovered the need to assert both the distinction and interdependence between units and relations. It now recognizes that, for an interaction to be real, the "nature" of the related things must derive from these relations, while at the same time the relations must derive from the "nature" of the things (see Chapter X). This is the forerunner of "self-consistent" descriptions as expressed, for instance, by the "bootstrap" philosophy in elementary-particle physics, which asserts the universal connectedness of all particles. However, when Whitehead wrote *Process and Reality,* the situation of physics was quite different, and Whitehead's philosophy found an echo only in biology.[23]

Whitehead's case as well as Bergson's convince us that only an opening, a widening of science can end the dichotomy between science and philosophy. This widening of science is possible only if we revise our conception of time. To deny time—that is, to reduce it to a mere deployment of a reversible law—is to abandon the possibility of defining a conception of nature coherent with the hypothesis that nature produced living beings, particularly man. It dooms us to choosing between an antiscientific philosophy and an alienating science.

"Ignoramus, Ignorabimus": The Positivist's Strain

Another method of overcoming the difficulties of classical rationality implied in classical science was to separate what was scientifically most fruitful from what is "true." This is another form of the Kantian cleavage. In his 1865 address "On the Goal of the Natural Sciences," Kirchoff stated that the ultimate goal of science is to reduce every phenomenon to motion, motion that in turn is described by theoretical mechanics. A similar statement was made by Helmholtz, a chemist, physician, physicist, and physiologist who dominated the German universities at the time when they were becoming the hub of European science. He stated: "the phenomena of nature are to be referred back to motions of material particles possessing unchangeable moving forces, which are dependent upon conditions of space alone."[24]

The aim of the natural sciences, therefore, was to reduce all observations to the laws formulated by Newton and extended by such illustrious physicists and mathematicians as Lagrange, Hamilton, and others. We were not to ask why these forces exist and enter Newton's equation. In any case, we could not "understand" matter or forces even if we used these concepts to formulate the laws of dynamics. The why, the basic nature of forces and masses, remains hidden from us. Du Bois Reymond, as we already mentioned, expressed concisely the limitations of our knowledge: "Ignoramus, ignorabimus." Science provides no access to the mysteries of the universe. What then is science?

We have already referred to Mach's influential view: Science is part of the Darwinian struggle for life. It helps us to organize our experience. It leads to an economy of thought. Mathematical laws are nothing more than conventions useful for summarizing the results of possible experiments. At the end of the nineteenth century, scientific positivism exercised a great intellectual appeal. In France it influenced the work of eminent thinkers such as Duhem and Poincaré.

One more step in the elimination of "contemptible metaphysics" and we come to the Vienna school. Here science is granted jurisdiction over all positive knowledge and philosophy needed to keep this positive knowledge in order. This meant a radical submission of all rational knowledge and questions to science. When Reichenbach, a distinguished neopositivist philosopher, wrote a book on the "direction of time," he stated:

> There is no other way to solve the problem of time than the way through physics. More than any other science, physics has been concerned with the nature of time. If time is objective the physicist must have discovered the fact. If there is Becoming, the physicist must know it; but if time is merely subjective and Being is timeless, the physicist must have been able to ignore time in his construction of reality and describe the world without the help of time. . . . It is a hopeless enterprise to search for the nature of time without studying physics. If there is a solution to the philosophical problem of time, it is written down in the equations of mathematical physics.[25]

Reichenbach's work is of great interest to anyone wishing to see what physics has to say on the subject of time, but it is not so much a book on the philosophy of nature as an account of the way in which the problem of time challenges scientists, not philosophers.

What then is the role of philosophy? It has often been said that philosophy should become the science of science. Philosophy's objective would then be to analyze the methods of science, to axiomatize and to clarify the concepts used. Such a role would make of the former "queen of sciences" something like their housemaid. Of course, there is the possibility that this clarification of concepts would permit further progress, that philosophy understood in this way would, through the use of other methods—logic, semantics—produce new knowledge comparable to that of science proper. It is this hope that sustains the "analytic philosophy" so prevalent in Anglo-American circles. We do not want to minimize the interest of such an inquiry. However, the problems that concern us here are quite different. We do not aim to clarify or axiomatize existing knowledge but rather to close some fundamental gaps in this knowledge.

A New Start

In the first part of this book we described, on the one hand, dialogue with nature that classical science made possible and, on the other, the precarious cultural position of science. Is there a way out? In this chapter we have discussed some attempts to reach alternative ways of knowledge. We have also considered the positivist view, which separates science from reality.

The moments of greatest excitement at scientific meetings very often occur when scientists discuss questions that are likely to have no practical utility whatsoever, no survival value—topics such as possible interpretations of quantum mechanics, or the role of the expanding universe in our concept of time. If the positivistic view, which reduces science to a symbolic calculus, was accepted, science would lose much of its appeal. Newton's synthesis between theoretical concepts and active knowledge would be shattered. We would be back to the situation familiar from the time of Greece and Rome, with an unbridgeable gap between technical, practical knowledge on one side and theoretical knowledge on the other.

For the ancients, nature was a source of wisdom. Medieval nature spoke of God. In modern times nature has become so silent that Kant considered that science and wisdom, science and truth, ought to be completely separated. We have been living with this dichotomy for the past two centuries. It is time for it to come to an end. As far as science is concerned, the time is ripe for this to happen. From our present perspective, the first step toward a possible reunification of knowledge was the discovery in the nineteenth century of the theory of heat, of the laws of thermodynamics. Thermodynamics appears as the first form of a "science of complexity." This is the science we now wish to describe, from its formulation to recent developments.

BOOK TWO

THE SCIENCE OF COMPLEXITY

CHAPTER IV

ENERGY AND THE INDUSTRIAL AGE

Heat, the Rival of Gravitation

Ignis mutat res. Ageless wisdom has always linked chemistry to the "science of fire." Fire became part of experimental science during the eighteenth century, starting a conceptual transformation that forced science to reconsider what it had previously rejected in the name of a mechanistic world view, topics such as irreversibility and complexity.

Fire transforms matter; fire leads to chemical reactions, to processes such as melting and evaporation. Fire makes fuel burn and release heat. Out of all this common knowledge, nineteenth-century science concentrated on the single fact that combustion produces heat and that heat may lead to an increase in volume; as a result, combustion produces work. Fire leads, therefore, to a new kind of machine, the heat engine, the technological innovation on which industrial society has been founded.[1]

It is interesting to note that Adam Smith was working on his *Wealth of Nations* and collecting data on the prospects and determinants of industrial growth at the same university at which James Watt was putting the finishing touches on his steam engine. Yet the only use for coal that Adam Smith could find was to provide heat for workers. In the eighteenth century, wind, water, and animals, and the simple machines driven by them, were still the only conceivable sources of power.

The rapid spread of the British steam engine brought about a new interest in the mechanical effect of heat, and thermodynamics, born out of this interest, was thus not so much concerned with the *nature* of heat as with heat's *possibilities* for producing "mechanical energy."

As for the birth of the "science of complexity," we propose to date it in 1811, the year Baron Jean-Joseph Fourier, the prefect of Isère, won the prize of the French Academy of Sciences for his mathematical description of the propagation of heat in solids.

The result stated by Fourier was surprisingly simple and elegant: heat flow is proportional to the gradient of temperature. It is remarkable that this simple law applies to matter, whether its state is solid, liquid, or gaseous. Moreover, it remains valid whatever the chemical composition of the body is, whether it is iron or gold. It is only the coefficient of proportionality between the heat flow and the gradient of temperature that is specific to each substance.

Obviously, the universal character of Fourier's law is not directly related to dynamic interactions as expressed by Newton's law, and its formulation may thus be considered the starting point of a new type of science. Indeed, the simplicity of Fourier's mathematical description of heat propagation stands in sharp contrast to the complexity of matter considered from the molecular point of view. A solid, a gas, or a liquid are macroscopic systems formed by an immense number of molecules, and yet heat conductivity is described by a single law. Fourier formulated his result at the time when Laplace's school dominated European science. Laplace, Lagrange, and their disciples vainly joined forces to criticize Fourier's theory, but they were forced to retreat.[2] At the peak of its glory, the Laplacian dream met with its first setback. A physical theory had been created that was every bit as mathematically rigorous as the mechanical laws of motion but that remained completely alien to the Newtonian world. From this time on, mathematics, physics, and Newtonian science ceased to be synonymous.

The formulation of the law of heat conduction had a lasting influence. Curiously, in France and Britain it was the starting point of different historical paths leading to our time.

In France, the failure of Laplace's dream led to the positivist classification of science into the well-defined compartments introduced by Auguste Comte. The Comtean division of science has been well analyzed by Michel Serrès[3]—heat and gravity, two universals, coexist in physics. Worse, as Comte was to state later, they are antagonistic. Gravitation acts on an inert mass that *submits* to it without being affected by it in any other way than by the motion it acquires or transmits. Heat *transforms* matter, determines changes of state, and leads to a modification of intrinsic properties. This was, in a sense, a confirmation of the protest made by the anti-Newtonian chemists of the eighteenth century and by all those who emphasized the difference between the purely spatiotemporal behavior attributed to mass and the specific activity of matter. This distinction was used as a foundation for the classification of the sciences, all placed by Comte under the common sign of order—that is, of equilibrium. To the mechanical equilibrium between forces the positivist classification simply adds the concept of thermal equilibrium.

In Britain, on the other hand, the theory of heat propagation did not mean giving up the attempt to unite the fields of knowledge but opened a new line of inquiry, the progressive formulation of a theory of irreversible processes.

Fourier's law, when applied to an isolated body with an unhomogeneous temperature distribution, describes the gradual onset of thermal equilibrium. The effect of heat propagation is to equalize progressively the distribution of temperature until homogeneity is reached. Everyone knew that this was an irreversible process; a century before, Boerhave had stressed that heat always spread and leveled out. The science of complex phenomena—involving interaction among a large number of particles—and the occurrence of temporal asymmetry thus were linked from the outset. But heat conduction did not become the starting point of an investigation into the nature of irreversibility before it was first linked with the notion of dissipation as seen from an engineering point of view.[4]

Let us go into some detail about the structure of the new "science of heat" as it took shape in the early nineteenth century. Like mechanics, the science of heat implied both an original conception of the physical *object* and a definition of machines or *engines*—that is, an identification of cause and effect in a specific mode of production of mechanical work.

The study of the physical processes involving heat entails defining a system, not, as in the case of dynamics, by the position and velocity of its constituents (there are some 10^{23} molecules in a volume of gas or a solid fragment of the order of a cm^3), but by *a set of macroscopic parameters* such as temperature, pressure, volume, and so on. In addition, we have to take into account the *boundary conditions* that describe the relation of the system to its environment.

Let us consider specific heat, one of the characteristic properties of a macroscopic system, as an example. The specific heat is a measure of the amount of heat required to raise the temperature of a system by one degree while its volume or pressure is kept constant. To study the specific heat—for instance, at constant volume—the system must be brought into interaction with its environment; it must receive a certain amount of heat while at the same time its volume is kept constant and its pressure is allowed to vary.

More generally, a system may be subjected to *mechanical* action (for example, either the pressure or the volume may be fixed by using a piston device), *thermal* action (a certain amount of heat may be given to or removed from the system, or the system itself may be brought to a given temperature through heat exchange), or *chemical* action (a flux of reactants and reaction products between the system and the environment). As we have already mentioned, pressure, volume, chemical composition, and temperature are the classical physicochemical parameters in terms of which the properties of macroscopic systems are defined. *Thermodynamics* is the science of the correlation among the variations in these properties. In comparison with dynamic objects, thermodynamic objects therefore lead to a new point of view. The aim of the theory is not to predict the changes in the system in terms of the interactions among particles; it aims instead to predict how the system will react to modifications we may impose on it from the outside.

A *mechanical* engine gives back in the form of work the potential energy it has received from the outside world. Both cause and effect are of the same nature and, at least ideally, equivalent. In contrast, the *heat* engine implies material changes of states, including the transformation of the system's mechanical properties, dilatation, and expansion. The mechanical work produced must be seen as the result of a true process of transformation and not only as a transmission of movement. Thus the heat engine is not merely a passive device; strictly speaking, it *produces* motion. This is the origin of a new problem: in order to *restore* the system's capacity to produce motion, the system must be brought back to its initial state. Thus a *second* process is needed, a second change of state that *compensates* for the change producing the motion. In a heat engine, this second process, which is opposite to the first, involves cooling the system until it regains its initial temperature, pressure, and volume.

The problem of the efficiency of heat engines, of the ratio between the work done and the heat that must be supplied to the system *to produce the two mutually compensating processes,* is the very point at which the concept of irreversible process was introduced into physics. We shall return to the importance of Fourier's law in this context. Let us first describe the essential role played by the principle of energy conservation.

The Principle of the Conservation of Energy

We have already emphasized the central place of energy in classical dynamics. The Hamiltonian (the sum of the kinetic and potential energies) is expressed in terms of canonical variables—coordinates and momenta—and leads to changes in these variables while itself remaining constant throughout the motion. Dynamic change merely modifies the respective importance of potential and kinetic energy, conserving their totality.

The early nineteenth century was characterized by unprecedented experimental ferment.[5] Physicists realized that motion does more than bring about changes in the relative position of bodies in space. New processes identified in the laboratories gradually formed a network that ultimately linked all the new fields of physics with other, more traditional branches, such as mechanics. One of these connections was accidentally discovered by Galvani. Before him, only static electric charges were known. Galvani, using a frog's body, set up the first experimental electric current. Volta soon recognized that the "galvanic" contractions in the frog were actually the effect of an electric *current* passing through it. In 1800, Volta constructed a chemical battery; electricity could thus be produced by chemical reactions. Then came electrolysis: electric current can modify chemical affinities and produce chemical reactions. But this current can also produce light and heat; and, in 1820, Oersted discovered the magnetic effects produced by electrical currents. In 1822, Seebeck showed that, inversely, heat could produce electricity and, in 1834, how matter could be cooled by electricity. Then, in 1831, Faraday induced an electric current by means of magnetic effects. A whole network of new effects was gradually uncovered. The scientific horizon was expanding at an unprecedented rate.

In 1847 a decisive step was taken by Joule: the links among chemistry, the science of heat, electricity, magnetism, and biology were recognized as a "*conversion.*" The idea of conversion, which postulates that "something" is quantitatively conserved while it is qualitatively transformed, generalizes what occurs during mechanical motion. As we have seen, total energy is conserved while potential energy is converted into kinetic energy, or vice versa. Joule defined a general *equivalent* for physicochemical transformations, thus making it possible to measure the quantity conserved. This quantity was later[6] to become known as "energy." He established the first equivalence by measuring the mechanical work required to raise the temperature of a given quantity of water by one degree. A unifying element had been discovered in the middle of a bewildering variety of new discoveries. The conservation of energy, throughout the various transformations undergone by physical, chemical, and biological systems, was to provide a guiding principle in the exploration of the new processes.

No wonder that the principle of the conservation of energy was so important to nineteenth-century physicists. For many of them it meant the unification of the whole of nature. Joule expressed this conviction in an English context:

> Indeed the phenomena of nature, whether mechanical, chemical, or vital, consist almost entirely in a continual conversion of attraction through space, living force (N.B., kinetic energy) and heat into one another. Thus it is that order is maintained in the universe—nothing is deranged, nothing ever lost, but the entire machinery, com-

> plicated as it is, works smoothly and harmoniously. And though, as in the awful vision of Ezekiel, "wheel may be in the middle of wheel," and everything may appear complicated and involved in the apparent confusion and intricacy of an almost endless variety of causes, effects, conversions, and arrangements, yet is the most perfect regularity preserved—the whole being governed by the sovereign will of God.[7]

The case of the Germans Helmholtz, Mayer, and Liebig—all three belonging to a culture that would have rejected Joule's conviction on the grounds of strictly positivist practice—is even more striking. At the time of their discoveries, none of the three was, strictly speaking, a physicist. On the other hand, all of them were interested in the physiology of respiration. This had become, since Lavoisier, a model problem in which the functioning of a living being could be described in precise physical and chemical terms, such as the combustion of oxygen, the release of heat, and muscular work. It was thus a question that would attract physiologists and chemists hostile to Romantic speculation and eager to contribute to experimental science. However, judging from the account of how these three scientists came to the conclusion that respiration, and then the whole of nature, was governed by some fundamental "equivalence," we may state that the German philosophic tradition had imbued them with a conception that was quite alien to a positivist position: without hesitation they all concluded that the whole of nature, in each of its details, is ruled by this single principle of conservation.

The case of Mayer is the most remarkable.[8] As a young doctor working in the Dutch colonies in Java, he noticed the bright red color of the venous blood of one of his patients. This led him to conclude that, in a warm, tropical climate, the inhabitants need to burn less oxygen to maintain body temperature; this results in the bright color of their blood. Mayer went on to establish the balance between oxygen consumption, which is the source of energy, and the energy consumption involved in maintaining body temperature despite heat losses and manual work. This was quite a leap, since the color of the blood could as well be due to the patient's "laziness." But Mayer went further and concluded that the balance between oxygen consumption and heat loss was merely the particular manifestation of the existence of an indestructible "force" underlying all phenomena.

This tendency to see natural phenomena as the products of an underlying reality that remains constant throughout its transformations is strikingly reminiscent of Kant. Kant's influence can also be recognized in another idea held by some physiologists, the distinction between vitalism as philosophical speculation and the problem of scientific methodology. For those physiologists, even if there was a "vital" force underlying the function of living beings, the *object* of physiology would nonetheless be purely physicochemical in nature. From the two points of view mentioned, Kantianism, which ratified the systematic form taken by mathematical physics during the eighteenth century, can also be identified as one of the roots of the renewal of physics in the nineteenth century.[9]

Helmholtz quite openly acknowledged Kant's influence. For Helmholtz, the principle of the conservation of energy was merely the embodiment in physics of the general a priori requirement on which all science is based—the postulate that there is a basic invariance underlying natural transformations:

> The problem of the sciences is, in the first place, to seek the laws by which the particular processes of nature may be referred to, and deduced from, general rules.
>
> We are justified, and indeed impelled in this proceeding, by the conviction that every change in nature must have a sufficient cause. The proximate causes to which we refer phenomena may, in themselves, be either variable or invariable; in the former case the above conviction impels us to seek for causes to account for the change, and thus we proceed until we at length arrive at final causes which are unchangeable, and which therefore must, in all cases where the exterior conditions are the same, produce the same invariable effects. The final aim of the theoretic natural sciences is therefore to discover the ultimate and unchangeable causes of natural phenomena.[10]

With the principle of the conservation of energy, the idea of a new golden age of physics began to take shape, an age that would lead to the ultimate generalization of mechanics.

The cultural implications were far-reaching, and they included a conception of society and men as energy-transforming engines. But energy conversion cannot be the whole story. It represents the aspects of nature that are peaceful and controllable, but below there must be another, more "active" level. Nietzsche was one of those who detected the echo of creations and destructions that go far beyond mere conservation or conversion. Indeed, only difference, such as a difference of temperature or of potential energy, can produce results that are also differences.[11] Energy conversion is merely the destruction of a difference, together with the creation of another difference. The power of nature is thus concealed by the use of equivalences. However, there is another aspect of nature that involves the boilers of steam engines, chemical transformations, life and death, and that goes beyond equivalences and conservation of energy.[12] Here we reach the most original contribution of thermodynamics, the concept of irreversibility.

Heat Engines and the Arrow of Time

When we compare mechanical devices to thermal engines, for example, to the red-hot boilers of locomotives, we can see at a glance the gap between the classical age and nineteenth-century technology. Still, physicists first thought that this gap could be ignored, that thermal engines could be described like mechanical ones, neglecting the crucial fact that fuel used by the steam engine disappears forever. But such complacency soon became impossible. For classical mechanics the symbol of nature was the clock; for the Industrial Age, it became a reservoir of energy that is always threatened with exhaustion. The world is burning like a furnace; energy, although being conserved, also is being dissipated.

The original formulation of the second law of thermodynamics, which would lead to the first quantitative expression of irreversibility, was made by Sadi Carnot in 1824, before the general formulation of the principle of conservation of energy by Mayer (1842) and Helmholtz (1847). Carnot analyzed the heat engine, closely following the work of his father, Lazare Carnot, who had produced an influential description of mechanical engines.

The description of mechanical engines assumes motion as a given. In modern language this corresponds to conservation of energy and momentum. Motion is merely converted and transferred to other bodies. But the analogy between mechanical and thermal engines was a natural one for Sadi Carnot, since he assumed, with most of the scientists of his time, that heat as well as mechanical energy are conserved.

Water falling from one level to another can drive a mill. Similarly, Sadi Carnot assumed two sources, one of which gives heat to the engine system, and the other, at a different temperature, which absorbs the heat given by the former. It is the *motion* of the heat through the engine, between the two sources at different temperatures—that is, the driving force of fire—that will make the engine work.

Carnot repeated his father's questions.[13] Which machine will have the highest efficiency? What are the sources of loss? What are the processes whereby heat propagates without producing work? Lazare Carnot had concluded that in order to obtain maximum efficiency from a mechanical machine it must be built and made to function to reduce to a minimum shocks, friction, or discontinuous changes of speed—in short, all that is caused by the sudden contact of bodies moving at different speeds. In doing so he had merely applied the physics of his time: only continuous phenomena are conservative; all abrupt changes in motion cause an irreversible loss of the "living force." Similarly, the ideal *heat engine*, instead of having to avoid all contacts between bodies moving at different *speeds*, will have to avoid all contact between bodies having different *temperatures*.

The cycle therefore has to be designed so that no temperature change results from direct heat flow between two bodies at different temperatures. Since such flows have no mechanical effect, they would merely lead to a loss of efficiency.

The ideal Carnot cycle is thus a rather tricky device that achieves the paradoxical result of a heat transfer between two sources at different temperatures without any contact between bodies of different temperatures. It is divided into four phases. During each of the two isothermal phases, the system is in contact with one of the two heat sources and is kept at the temperature of this source. When in contact with the hot source, it absorbs heat and expands; when in contact with the cold source, it loses heat and contracts. The two isothermal phases are linked up by two phases in which the system is isolated from the sources—that is, heat no longer enters or leaves the system, but the temperature of the latter changes as a result, respectively, of expansion and compression. The volume continues to change until the system has passed from the temperature of one source to that of the other.

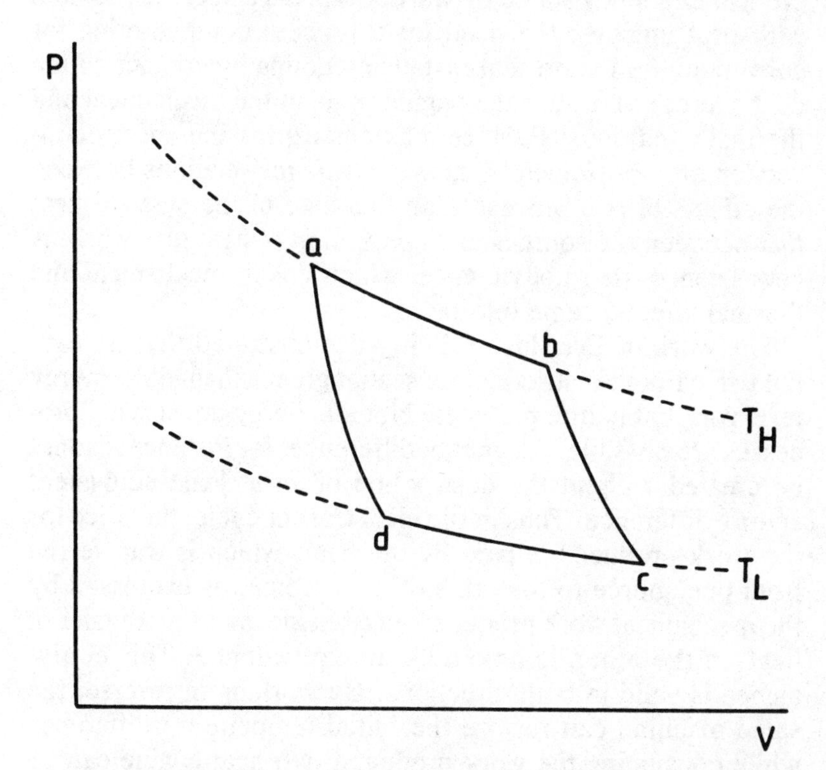

Figure 2. Pressure-volume diagram of the Carnot cycle: a thermodynamic engine, functioning between two sources, one "hot" at temperature T_H, the other "cold" at temperature T_L. Between state *a* and state *b*, there is an isothermal change: The system, kept at temperature T_H, absorbs heat and expands. Between *b* and *c*, the system is kept expanding while in thermal isolation; its temperature goes down from T_H to T_L. Those two steps produce mechanical energy. Between *c* and *d*, there is a second isothermal change: the system is compressed and releases heat while being kept at temperature T_L. Between *d* and *a*, the system, again isolated, is compressed while its temperature increases to temperature T_H.

It is quite remarkable that this description of an ideal thermal engine does not mention the irreversible processes that are at the basis of its realization. No mention is made of the furnace in which the coal is burning. The model is only concerned with the *effect* of the combustion, which permits the maintenance of the temperature difference between the two sources.

In 1850, Clausius described the Carnot cycle from the new perspective provided by the conservation of energy. He discovered that the need for two sources and the formula for theoretical efficiency stated by Carnot express a specific problem with heat engines: the need for a process compensating for conversion (in the present instance, cooling by contact with a cold source) to restore the engine to its initial mechanical and thermal conditions. Balance relations expressing energy conversion are now joined by new *equivalence* relations between the effects of two processes on the state of the system, heat *flux* between the sources and *conversion* of heat into work. A new science, thermodynamics, which linked mechanical and thermal effects, came into being.

The work of Clausius explicitly demonstrated that we cannot use without restriction the seemingly inexhaustible energy reservoir that nature provides. Not all energy-conserving processes are possible. An energy difference, for instance, cannot be created without the destruction of an at least equivalent energy difference. Thus in the ideal Carnot cycle, the price for the work produced is paid by the heat, which is transferred from one source to the other. The outcome, as expressed by the mechanical work produced on one side, and the transfer of heat on the other, is linked by an equivalence. This equivalence is valid in both directions. By working in reverse, the same machine can restore the initial temperature difference while consuming the work produced. No heat engine can be constructed using a single source of heat.

Clausius was no more concerned than Carnot with the losses whereby all real engines have an efficiency lower than the ideal value predicted by the theory. His description, like that of Carnot, corresponds to an idealization. It leads to the definition of the limit nature imposes on the yield of thermal engines.

However, since the eighteenth century, the status of idealizations had changed. Based as it was on the principle of the conservation of energy, the new science claimed to describe not only idealizations, but also nature itself, *including "losses."* This raised a new problem, whereby irreversibility entered physics. How does one describe what happens in a real engine? How does one include losses in the energy balance? How do they reduce efficiency? These questions paved the way to the second law of thermodynamics.

From Technology to Cosmology

As we have seen, the question raised by Carnot and Clausius led to a description of ideal engines that was based on conservation and compensation. In addition, it provided an opportunity for presenting new problems, such as the dissipation of energy. William Thomson, who had great respect for Fourier's work, was quick to grasp the importance of the problem, and in 1852 he was the first to formulate the second law of thermodynamics.

It was Fourier's heat propagation that Carnot had identified as a possible cause for the power losses in a heat engine. Carnot's cycle, no longer the ideal cycle but the "real" cycle, thus became the point of convergence of the *two universalities* discovered in the nineteenth century—energy conversion and heat propagation. The combination of these two discoveries led Thomson to formulate his new principle: the existence in nature of a *universal tendency* toward the degradation of mechanical energy. Note the word "universal," which has obvious cosmological connotations.

The world of Laplace was eternal, an ideal perpetual-motion machine. Since Thomson's cosmology is not merely a reflection of the new ideal heat engine but also incorporates the consequences of the irreversible propagation of heat in a world in which energy is conserved. This world is described as an engine in which heat is converted into motion only at the price of some irreversible waste and useless dissipation. Effect-producing differences in nature progressively diminish. The world uses up its differences as it goes from one conversion to another and tends toward a final state of thermal equilibrium, "heat death." In accordance with Fourier's law, in the end there will no longer be any differences of temperature to produce a mechanical effect.

Thomson thus made a dizzy leap from engine technology to cosmology. His formulation of the second law was couched in the scientific terminology of his time: the conservation of energy, engines, and Fourier's law. It is clear, moreover, that the part played by the cultural context was important. It is generally accepted that the problem of time took on a new importance during the nineteenth century. Indeed, the essential role of time began to be noticed in all fields—in geology, in biology, in language, as well as in the study of human social evolution and ethics. But it is interesting that the specific form in which time was introduced in physics, as a tendency toward homogeneity and death, reminds us more of ancient mythological and religious archetypes than of the progressive complexification and diversification described by biology and the social sciences. The return of these ancient themes can be seen as a cultural repercussion of the social and economic upheavals of the time. The rapid transformation of the technological mode of interaction with nature, the constantly accelerating pace of change experienced by the nineteenth century, produced a deep anxiety. This anxiety is still with us and takes various forms, from the repeated proposals for a "zero growth" society or for a moratorium on scientific research to the announcement of "scientific truths" concerning our disintegrating universe. Present knowledge in astrophysics is still scanty and very problematic, since in this field gravitational effects play an essential role and problems imply the simultaneous use of thermodynamics and relativity. Yet most texts in this field are unanimous in predicting final doom. The conclusion of a recent book reads:

> The unpalatable truth appears to be that the inexorable disintegration of the universe as we know it seems assured, the organization which sustains all ordered activity, from men to galaxies, is slowly but inevitably running down, and may even be overtaken by total gravitational collapse into oblivion.[14]

Others are more optimistic. In an excellent article on the energy of the universe, Freeman Dyson has written:

> It is conceivable however that life may have a larger role to play than we have yet imagined. Life may succeed against all of the odds in molding the universe to its own purpose. And the design of the inanimate universe may not be as detached from the potentialities of life and intelligence as scientists of the twentieth century have tended to suppose.[15]

In spite of the important progress made by Hawking and others, our knowledge of large-scale transformations in our universe remains inadequate.

The Birth of Entropy

In 1865, it was Clausius' turn to make the leap from technology to cosmology. At the outset he merely reformulated his earlier conclusions, but in doing so he introduced a new concept, *entropy.* His first goal was to distinguish clearly between the concepts of conservation and of reversibility. Unlike mechanical transformations, where reversibility and conservation coincide, a physicochemical transformation may conserve energy even though it cannot be reversed. This is true, for instance, in the case of friction, in which motion is converted into heat, or in the case of heat conduction as it was described by Fourier.

We are already familiar with energy, which is a function of the state of a system—that is, a function dependent only on the value of the parameters (pressure, volume, temperature) by which that state may be defined.[16] But we must go beyond the principle of energy conservation and find a way to express the distinction between "useful" exchanges of energy in the Carnot cycle and "dissipated" energy that is irreversibly wasted.

This is precisely the role of Clausius' new function, entropy, generally denoted by S.

Apparently Clausius merely wished to express in a new form

the obvious requirement that an engine return to its initial state at the end of its cycle. The first definition of entropy is centered on conservation: at the end of each cycle, whether ideal or not, the function of the system's state, entropy, returns to its initial value. But the parallel between entropy and energy ends as soon as we abandon idealizations.[17]

Let us consider the variation of the entropy dS over a short time interval dt. The situation is quite different for ideal and real engines. In the first case, dS may be expressed completely in terms of the exchanges between the engine and its environment. We can set up experiments in which heat is given up by the system instead of flowing into the system. The corresponding change in entropy would simply have its sign changed. This kind of contribution to entropy, which we shall call d_eS, is therefore *reversible* in the sense that it can have either a positive or a negative sign. The situation is drastically different in a real engine. Here, in addition to reversible exchanges, we have *irreversible* processes inside the system, such as heat losses, friction, and so on. These produce an entropy increase or "entropy production" inside the system. This increase of entropy, which we shall call d_iS, cannot change its sign through a reversal of the heat exchange with the outside world. Like all irreversible processes (such as heat conduction), entropy production always proceeds in the same direction. In other words, d_iS can only be positive or vanish in the absence of irreversible processes. Note that the positive sign of d_iS is chosen merely by convention; it could just as well have been negative. The point is that the variation is monotonous, that entropy production cannot change its sign as time goes on.

The notations d_eS and d_iS have been chosen to remind the reader that the first term refers to *exchanges* (*e*) with the outside world, while the second refers to the irreversible processes *inside* (*i*) the system. The entropy variation dS is therefore the sum of the two terms d_eS and d_iS, which have quite different physical meanings.[18]

To grasp the peculiar feature of this decomposition of entropy variation into two parts, it is useful to apply our formulation to energy. Let us denote energy by E and variation over a short time dt by dE. Of course, we would still write that dE is equal to the sum of a term d_eE due to the exchanges of energy and a term d_iE linked to the "internal production" of energy. However, the principle of the conservation of energy states that energy is never "produced" but only transferred from one place to another. The variation in energy dE is then reduced to d_eE. On the other hand, if we take a nonconserved quantity, such as the quantity of hydrogen molecules contained in a vessel, this quantity may vary both as the result of adding hydrogen to the vessel or through chemical reactions occurring inside the vessel. But in this case, the sign of the "production" is not determined. Depending on the circumstances, we can produce or destroy hydrogen molecules by transferring hydrogen atoms to other chemical components. The peculiar feature of the second law is the fact that the production term d_iS is *always positive.* The production of entropy expresses the occurrence of irreversible changes inside the system.

Clausius was able to express quantitatively the entropy flow d_eS in terms of the heat received (or given up) by the system. In a world dominated by the concepts of reversibility and conservation, this was his main concern. Regarding the irreversible processes involved in entropy production, he merely stated the existence of the inequality $d_iS/dt>0$. Even so, important progress had been made, for, if we leave the Carnot cycle and consider other thermodynamic systems, the *distinction between entropy flow and entropy production can still be made.* For an isolated system that has no exchanges with its environment, the entropy flow is, by definition, zero. Only the production term remains, and the system's entropy can only increase or remain constant. Here, then, it is no longer a question of irreversible transformations considered as approximations of reversible transformations; increasing entropy corresponds to the *spontaneous evolution* of the system. Entropy thus becomes an "indicator of evolution," or an "arrow of time," as Eddington aptly called it. For all isolated systems, the future is the direction of increasing entropy.

What system would be better "isolated" than the universe as a whole? This concept is the basis of the cosmological formulation of the two laws of thermodynamics given by Clausius in 1865:

Die Energie der Welt ist konstant.

Die Entropie der Welt strebt einem Maximum zu.[19]

The statement that the entropy of an isolated system increases to a maximum goes far beyond the technological problem that gave rise to thermodynamics. Increasing entropy is no longer synonymous with loss but now refers to the *natural processes* within the system. These are the processes that ultimately lead the system to thermodynamic "equilibrium" corresponding to the state of maximum entropy.

In Chapter I we emphasized the element of surprise involved in the discovery of Newton's universal laws of dynamics. Here also the element of surprise is apparent. When Sadi Carnot formulated the laws of ideal thermal engines, he was far from imagining that his work would lead to a conceptual revolution in physics.

Reversible transformations belong to classical science in the sense that they define the possibility of acting on a system, of controlling it. The *dynamic object* could be controlled through its initial conditions. Similarly, when defined in terms of its reversible transformations, the *thermodynamic object* may be controlled through its boundary conditions: any system in thermodynamic equilibrium whose temperature, volume, or pressure are *gradually* changed passes through a series of equilibrium states, and any reversal of the manipulation leads to a return to its initial state. The reversible nature of such change and controlling the object through its boundary conditions are interdependent processes. In this context irreversibility is "negative"; it appears in the form of "uncontrolled" changes that occur as soon as the system eludes control. But inversely, irreversible processes may be considered as the last remnants of the spontaneous and intrinsic *activity* displayed by nature when experimental devices are employed to harness it.

Thus the "negative" property of dissipation shows that, unlike dynamic objects, thermodynamic objects can only be *partially* controlled. Occasionally they "break loose" into spontaneous change.

All changes are not equivalent for a thermodynamic system. This is the meaning of the expression $dS = d_eS + d_iS$. Spontaneous change toward equilibrium d_iS is different from the change d_eS, which is determined and controlled by a modification of the boundary conditions (for example, ambient temperature). For an isolated system, equilibrium appears as an "attractor" of nonequilibrium states. Our initial assertion may thus be generalized by saying that evolution toward an attractor state differs from all other changes, especially from changes determined by boundary conditions.

Max Planck often emphasized the difference between the two types of change found in nature. Nature, wrote Planck, seems to "favor" certain states. The irreversible increase in entropy d_iS/dt describes a system's approach to a state which "attracts" it, which the system prefers and from which it will not move of its own "free will." "From this point of view, Nature does not permit processes whose final states she finds less attractive than their initial states. Reversible processes are limiting cases. In them, Nature has an equal propensity for initial and final states; this is why the passage between them can be made in both directions."[20]

How foreign such language sounds when compared with the language of dynamics! In dynamics, a system changes according to a trajectory that is given once and for all, whose starting point is never forgotten (since initial conditions determine the trajectory for all time). However, in an isolated system *all* nonequilibrium situations produce evolution toward the *same* kind of equilibrium state. By the time equilibrium has been reached, the system has *forgotten* its initial conditions—that is, the way it had been prepared.

Thus specific heat or the compressibility of a system in equilibrium are properties independent of the way the system has been set up. This fortunate circumstance greatly simplifies the study of the physical states of matter. Indeed, complex systems consist of an immense number of particles.* From the dynamic standpoint it is practically impossible to reproduce any state of such systems in view of the infinite variety of dynamic states that may occur.

We are now confronted with two basically different descriptions: dynamics, which applies to the world of motion, and

*Physical chemistry often employs Avogadro's number—that is, the number of molecules in a "mole" of matter (a mole always contains the same number of particles, the number of atoms contained in one gram of hydrogen). This number is of the order of 6.10^{23}, and it is the characteristic order of magnitude of the number of particles forming systems governed by the laws of classical thermodynamics.

thermodynamics, the science of complex systems with its intrinsic direction of evolution toward increasing entropy. This dichotomy immediately raises the question of how these descriptions are related, a problem that has been debated since the laws of thermodynamics were formulated.

Boltzmann's Order Principle

The second law of thermodynamics contains two fundamental elements: (1) a "negative" one that expresses the impossibility of certain processes (heat flows from the hot source to the cold and not vice versa) and (2) a "positive," constructive one. The second is a consequence of the first; it is the impossibility of certain processes that permits us to introduce a function, entropy, which increases uniformly for isolated systems. Entropy behaves as an attractor for isolated systems.

How could the formulations of thermodynamics be reconciled with dynamics? At the end of the nineteenth century, most scientists seemed to think this was impossible. The principles of thermodynamics were new laws forming the basis of a new science that could not be reduced to traditional physics. Both the qualitative diversity of energy and its tendency toward dissipation had to be accepted as new axioms. This was the argument of the "energeticists" as opposed to the "atomists," who refused to abandon what they considered to be the essential mission of physics—to reduce the complexity of natural phenomena to the simplicity of elementary behavior expressed by the laws of motion.

Actually, the problems of the transition from the microscopic to the macroscopic level were to prove exceptionally fruitful for the development of physics as a whole. Boltzmann was the first to take up the challenge. He felt that new concepts had to be developed to extend the physics of trajectories to cover the situation described by thermodynamics. Following in Maxwell's footsteps, Boltzmann sought this conceptual innovation in the theory of probability.

That probability could play a role in the description of complex phenomena was not surprising: Maxwell himself appears to have been influenced by the work of Quetelet, the inventor of the "average" man in sociology. The innovation was to introduce probability in physics not as a means of approximation but rather as an explanatory principle, to use it to show that a system could display a new type of behavior by virtue of its being composed of a large population to which the laws of probability could be applied.

Let us consider a simple example of the application of the concept of probability in physics. An ensemble composed of N particles is contained in a box divided into two equal compartments. The problem is to find the probability of the various possible distributions of particles between the compartments—that is, the probability of finding N_1 particles in the first compartment (and $N_2 = N - N_1$ in the second).

Using combinatorial analysis, it is easy to calculate the number of ways in which each different distribution of N particles can be achieved. Thus if $N = 8$, there is only one way of placing the eight particles in a single half. There are, however, eight different ways of putting one particle in one half and seven in the other half, if we suppose the particles to be distinguishable, as is assumed in classical physics. Furthermore, equal distribution of the eight particles between the two halves can be carried out in $8!/4!4! = 70$ different ways (where $n! = 1 \cdot 2 \cdot 3 \ldots (n-1) \cdot n$). Likewise, whatever the value of N, a number P of situations called *complexions* in physics may be defined, giving the number of ways of achieving any given distribution N_1, N_2. Its expression is $P = N!/N_1!N_2!$.

For any given population, the larger the number of complexions the smaller the difference between N_1 and N_2. It is maximum when the population is equally distributed over the two halves. Moreover, the larger the value of N, the greater the difference between the number of complexions corresponding to the different ways of distribution. For values of N of the order of 10^{23} values found in macroscopic systems, the overwhelming majority of possible distributions corresponds to the distribution $N_1 = N_2 = N/2$. For systems composed of a large number of particles, all states that differ from the state corresponding to an equal distribution are thus highly improbable.

Boltzmann was the first to realize that irreversible increase in entropy could be considered as the expression of a growing molecular disorder, of the gradual forgetting of any initial dissymmetry, since dissymmetry decreases the number of complexions when compared to the state corresponding to the maximum of *P*. Boltzmann thus aimed to identify entropy *S* with the number of complexions: entropy characterizes each macroscopic state in terms of the number of ways of achieving this state. Boltzmann's famous equation $S = k \, lg \, P$† expresses this idea in quantitative form. The proportionality factor *k* in this formula is a universal constant, known as Boltzmann's constant.

Boltzmann's results signify that irreversible thermodynamic change is a change toward states of increasing probability and that the attractor state is a macroscopic state corresponding to maximum probability. This takes us far beyond Newton. For the first time a physical concept has been explained in terms of probability. Its utility is immediately apparent. Probability can adequately explain a system's forgetting of all initial dissymmetry, of all special distributions (for example, the set of particles concentrated in a subregion of the system, or the distribution of velocities that is created when two gases of different temperatures are mixed). This forgetting is possible because, *whatever the evolution peculiar to the system,* it will ultimately lead to one of the microscopic states corresponding to the macroscopic state of disorder and maximum symmetry, since these macroscopic states correspond to the overwhelming majority of possible microscopic states. Once this state has been reached, the system will move only short distances from the state, and for short periods of time. In other words, the system will merely *fluctuate* around the attractor state.

Boltzmann's order principle implies that the most probable state available to a system is the one in which the multitude of events taking place simultaneously in the system *compensates for one another statistically.* In the case of our first example, whatever the initial distribution, the system's evolution will ultimately lead it to the equal distribution $N_1 = N_2$. This state will put an end to the system's irreversible macroscopic evolu-

---

†The logarithmic expression indicates that entropy is an additive quantity ($S_{1+2} = S_1 + S_2$), while the number of complexions is multiplicative ($P_{1+2} = P_1 \cdot P_2$).

tion. Of course, the particles will go on moving from one half to the other, but on the average, at any given instant, *as many will be going in one direction as in the other.* As a result, their motion will cause only small, short-lived fluctuations around the equilibrium state $N_1 = N_2$. Boltzmann's probabilistic interpretation thus makes it possible to understand the specificity of the attractor studied by equilibrium thermodynamics.

This is not the whole story, and we shall devote the third part of this book to a more detailed discussion. A few remarks suffice here. In classical mechanics (and, as we shall see, in quantum mechanics as well), everything is determined in terms of initial states and the laws of motion. How then does probability enter the description of nature? Here it is common to invoke our ignorance of the exact dynamic state of the system. This is the subjectivistic interpretation of entropy. Such an interpretation was acceptable when irreversible processes were considered to be mere nuisances corresponding to friction or, more generally, to losses in the functioning of thermal engines. But today the situation has changed. As we shall see, irreversible processes have an immense constructive importance: life would not be possible without them. The subjectivistic interpretation is therefore highly questionable. Are we ourselves merely the result of our ignorance, of the fact that we only observe macroscopic states.

Moreover, both in thermodynamics as well as in its probabilistic interpretation, there appears a dissymmetry in time: entropy increases in the direction of the future, not of the past. This seems impossible when we consider dynamic equations that are invariant in respect to time inversion. As we shall see, the second law is a selection principle compatible with dynamics but not deducible from it. It limits the possible initial conditions available to a dynamic system. The second law therefore marks a radical departure from the mechanistic world of classical or quantum dynamics. Let us now return to Boltzmann's work.

So far we have discussed isolated systems in which the number of particles as well as the total energy are fixed by the boundary conditions. However, it is possible to extend Boltzmann's explanation to open systems that interact with their environment. In a closed system, defined by boundary conditions such that its temperature T is kept constant by heat exchange with the environment, equilibrium is not defined in terms of maximum entropy but in terms of the minimum of a similar function, free energy: $F=E-TS$, where E is the energy of the system and T is the temperature (measured on the so-called Kelvin scale, where the freezing point of water is 273°K and its boiling point is 373°K).

This formula signifies that equilibrium is the result of competition between energy and entropy. Temperature is what determines the relative weight of the two factors. At low temperatures, energy prevails, and we have the formation of *ordered* (weak-entropy) and *low-energy* structures such as crystals. Inside these structures each molecule interacts with its neighbors, and the kinetic energy involved is small compared with the potential energy that results from the interactions of each molecule with its neighbors. We can imagine each particle as imprisoned by its interactions with its neighbors. At high temperatures, however, entropy is dominant and so is molecular disorder. The importance of relative motion increases, and the regularity of the crystal is disrupted; as the temperature increases, we first have the liquid state, then the gaseous state.

The entropy S of an isolated system and the free energy F of a system at fixed temperature are examples of "thermodynamic potentials." The extremes of thermodynamic potentials such as S or F define the attractor states toward which systems whose boundary conditions correspond to the definition of these potentials tend spontaneously.

Boltzmann's principle can also be used to study the coexistence of structures (such as the liquid phase and the solid phase) or the equilibrium between a crystallized product and the same product in solution. It is important to remember, however, that equilibrium structures are defined on the molecular level. It is the interaction between molecules acting over a range of the order of some 10^{-8} cm, the same order of magnitude as the diameter of atoms in molecules, that makes a crystal structure stable and endows it with its macroscopic properties. Crystal size, on the other hand, is not an intrinsic property of structure. It depends on the quantity of matter in the crystalline phase at equilibrium.

Carnot and Darwin

Equilibrium thermodynamics provides a satisfactory explanation for a vast number of physicochemical phenomena. Yet it may be asked whether the concept of equilibrium structures encompasses the different structures we encounter in nature. Obviously the answer is no.

Equilibrium structures can be seen as the results of statistical compensation for the activity of microscopic elements (molecules, atoms). By definition they are inert at the global level. For this reason they are also "immortal." Once they have been formed, they may be isolated and maintained indefinitely without further interaction with their environment. When we examine a biological cell or a city, however, the situation is quite different: not only are these systems open, but also they exist only because they are open. They feed on the flux of matter and energy coming to them from the outside world. We can isolate a crystal, but cities and cells die when cut off from their environment. They form an integral part of the world from which they draw sustenance, and they cannot be separated from the fluxes that they incessantly transform.

However, it is not only living nature that is profoundly alien to the models of thermodynamic equilibrium. Hydrodynamics and chemical reactions usually involve exchanges of matter and energy with the outside world.

It is difficult to see how Boltzmann's order principle can be applied to such situations. The fact that a system becomes more uniform in the course of time can be understood in terms of complexions; in a state of uniformity, when the "differences" created by the initial conditions have been forgotten, the number of complexions will be maximum. But it is impossible to understand spontaneous convection from this point of view. The convection current calls for coherence, for the cooperation of a vast number of molecules. It is the opposite of disorder, a privileged state to which only a comparatively small number of complexions may correspond. In Boltzmann's terms, it is an "improbable" state. If convection must be considered a "miracle," what then is there to say about life, with its highly specific features present in the simplest organisms?

The question of the relevance of equilibrium models can be reversed. In order to produce equilibrium, a system must be "protected" from the fluxes that compose nature. It must be "canned," so to speak, or put in a bottle, like the homunculus in Goethe's *Faust,* who addresses to the alchemist who created him: "Come, press me tenderly to your breast, but not too hard, for fear the glass might break. This is the way things are: something natural, the whole world hardly suffices what is, but what is artificial demands a closed space." In the world that we are familiar with, equilibrium is a rare and precarious state. Even evolution toward equilibrium implies a world like ours, far enough away from the sun for the partial isolation of a system to be conceivable (no "canning" is possible at the temperature of the sun), but a world in which nonequilibrium remains the rule, a "lukewarm" world where equilibrium and nonequilibrium coexist.

For a long time, however, physicists thought they could define the inert structure of crystals as the only physical order that is predictable and reproducible and approach equilibrium as the only evolution that could be deduced from the fundamental laws of physics. Thus any attempt at extrapolation from thermodynamic descriptions was to define as rare and unpredictable the kind of evolution described by biology and the social sciences. How, for example, could Darwinian evolution—the statistical *selection* of rare events—be reconciled with the statistical disappearance of all peculiarities, of all rare configurations, described by Boltzmann? As Roger Caillois[21] asks: "Can Carnot and Darwin both be right?"

It is interesting to note how similar in essence the Darwinian approach is to the path explored by Boltzmann. This may be more than a coincidence. We know that Boltzmann had immense admiration for Darwin. Darwin's theory begins with an assumption of the spontaneous fluctuations of species; then selection leads to irreversible biological evolution. Therefore, as with Boltzmann, a randomness leads to irreversibility. Yet the result is very different. Boltzmann's interpretation implies the forgetting of initial conditions, the "destruction" of initial structures, while Darwinian evolution is associated with self-organization, ever-increasing complexity.

To sum up our argument so far, equilibrium thermodynamics was the first response of physics to the problem of nature's complexity. This response was expressed in terms of the dissipation of energy, the forgetting of initial conditions, and evolution toward disorder. Classical dynamics, the science of eternal, reversible trajectories, was alien to the problems facing the nineteenth century, which was dominated by the concept of evolution. Equilibrium thermodynamics was in a position to oppose its view of time to that of other sciences: for thermodynamics, time implies degradation and death. As we have seen, Diderot had already asked the question: Where do we, organized beings endowed with sensations, fit in an inert world subject to dynamics? There is another question, which has plagued us for more than a century: What significance does the evolution of a living being have in the world described by thermodynamics, a world of ever-increasing disorder? What is the relationship between thermodynamic time, a time headed toward equilibrium, and the time in which evolution toward increasing complexity is occurring?

Was Bergson right? Is time the very medium of innovation, or is it nothing at all?

CHAPTER V

THE THREE STAGES OF THERMODYNAMICS

Flux and Force

Let us return[1] to the description of the second law given in the previous chapter. The concept of entropy plays a central role in the description of evolution. As we have seen, its variation can be written as the sum of two terms—the term d_eS, linked to the exchanges between the system and the rest of the world, and a production term, d_iS, resulting from irreversible phenomena inside the system. This term is always positive except at thermodynamic equilibrium, when it becomes zero. For isolated systems ($d_eS=0$), the equilibrium state corresponds to a state of maximum entropy.

In order to appreciate the significance of the second law for physics, we need a more detailed description of the various irreversible phenomena involved in the entropy production d_iS or in the entropy production per unit time $P=d_iS/dt$.

For us chemical reactions are of particular significance. Together with heat conduction, they form the prototype of irreversible processes. In addition to their intrinsic importance, chemical processes play a fundamental role in biology. The living cell presents an incessant metabolic activity. There thousands of chemical reactions take place simultaneously to transform the matter the cell feeds on, to synthesize the fundamental biomolecules, and to eliminate waste products. As regards both the different reaction rates and the reaction sites within the cell, this chemical activity is highly coordinated. The biological structure thus combines order and activity. In contrast, an equilibrium state remains inert even though it may be structured, as, for example, with a crystal. Can chemical processes provide us with the key to the difference between the behavior of a crystal and that of a cell?

We will have to consider chemical reactions from a dual point of view, both kinetic and thermodynamic.

From the kinetic point of view, the fundamental quantity is the reaction *rate*. The classical theory of chemical kinetics is based on the assumption that the rate of a chemical reaction is proportional to the concentrations of the reactants taking part in it. Indeed, it is through collisions between molecules that a reaction takes place, and it is quite natural to assume that the number of collisions is proportional to the product of the concentrations of the reacting molecules.

For the sake of example, let us take a simple reaction such as $A + X \rightarrow B + Y$. This "reaction equation" means that whenever a molecule of component A encounters a molecule of X, there is a certain probability that a reaction will take place and a molecule of B and a molecule of Y will be produced. A collision producing such a change in the molecules involved is a "reactive collision." Only a usually very small fraction (for example, $1/10^6$) of all collisions are of this kind. In most cases, the molecules retain their original nature and merely exchange energy.

Chemical kinetics deals with changes in the concentration of the different products involved in a reaction. This kinetics is described by differential equations, just as motion is described by the Newtonian equations. However, in this case, we are not calculating acceleration but the rates of change of concentration, and these rates are expressed as a function of the concentrations of the reactants. The rate of change of concentration of X, dX/dt, is thus proportional to the product of the concentrations of A and X in the solution—that is, $dX/dt = -kA \cdot X$, where k is a proportionality factor that is linked to quantities such as temperature and pressure and that provides a measure for the fraction of reactive collisions taking place and leading to the reaction $A + X \rightarrow Y + B$. Since, in the example taken, whenever a molecule of X disappears, a molecule of A disappears too, and a molecule of Y and one of B are formed, the rates of change of concentration are related: $dX/dt = dA/dt = -dY/dt = -dB/dt$.

But if the collision between a molecule of X and a molecule of A can set off a chemical reaction, the collision between molecules of Y and B can set off the opposite reaction. A second reaction $Y+B \rightarrow X+A$ thus occurs within the system described, bringing about a supplementary variation in the concentration of X, $dX/dt = k'YB$. The total variation in concentration of a chemical compound is given by the balance between the forward and the reverse reaction. In our example, $dX/dt \; (= -dY/dt = \ldots) = -kAX + k'YB$.

If left to itself, a system in which chemical reactions occur tends toward a state of chemical equilibrium. Chemical equilibrium is therefore a typical example of an "attractor" state. Whatever its initial chemical composition, the system spontaneously reaches this final stage, where the forward and reverse reactions compensate one another statistically so that there is no longer any overall variation in the concentrations ($dX/dt = 0$). This compensation implies that the ratio between equilibrium concentrations is given by $AX/YB = k'/k = K$. This result is known as the "law of mass action," or Guldberg and Waage's law, and K is the equilibrium constant. The ratio between concentrations determined by the law of mass action corresponds to chemical equilibrium in the same way that uniformity of temperature (in the case of an isolated system) corresponds to thermal equilibrium. The corresponding entropy production vanishes.

Before we deal with the thermodynamic description of chemical reactions, let us briefly consider an additional aspect of the kinetic description. The rate of chemical reactions is affected not only by the concentrations of the reacting molecules and thermodynamic parameters (for example, pressure and temperature) but also may be affected by the presence in the system of chemical substances that modify the reaction rate without themselves being changed in the process. Substances of this kind are known as "catalysts." Catalysts can, for instance, modify the value of the kinetic constants k or k' or even allow the system to follow a new "reaction path." In biology, this role is played by specific proteins, the "enzymes." These macromolecules have a spatial configuration that allows them to modify the rate of a given reaction. Often they are highly specific and affect only one reaction. A possible mechanism for the catalytic effect of enzymes is to present different "reaction sites" to which the different molecules involved in the reaction tend to attach themselves, thus increasing the likelihood of their coming into contact and reacting.

One very important type of catalysis, particularly in biology, is the one in which the presence of a product is required for its own synthesis. In other words, in order to produce the molecule X we must begin with a system already containing X. Very frequently, for instance, the molecule X *activates* an enzyme. By attaching itself to the enzyme it stabilizes that particular configuration in which the reaction site is available. To such an autocatalysis process correspond reaction schemes such as $A+2X \rightarrow 3X$; in the presence of molecules X, a molecule A is converted into a molecule X. Therefore we need X to produce more X. This reaction may be symbolized by the reaction "loop":

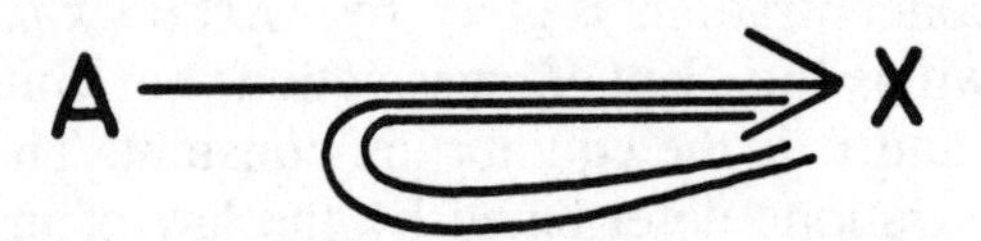

One important feature of systems involving such "reaction loops" is that the kinetic equations describing the changes occurring in them are *nonlinear* differential equations.

If we apply the same method as above, the kinetic equation obtained for the reaction $A+2X \rightarrow 3X$ *is* $dX/dt = kAX^2$, where the rate of variation of the concentration of X is proportional to the *square* of its concentration.

Another very important class of catalytic reactions in biology is that of crosscatalysis—for example, $2X+Y \rightarrow 3X$, $B+X \rightarrow Y+D$, which may be represented by the loop of Figure 3.

This is a case of crosscatalysis, since X is produced from Y, and simultaneously Y from X. Catalysis does not necessarily increase the reaction rate; it may, on the contrary, lead to inhibition, which can also be represented by suitable feedback loops.

The peculiar mathematical properties of the nonlinear differential equations describing chemical processes with catalytic steps are vitally important, as we shall see later, for the thermodynamics of far-from-equilibrium chemical processes. In addition, as we have already mentioned, molecular biology

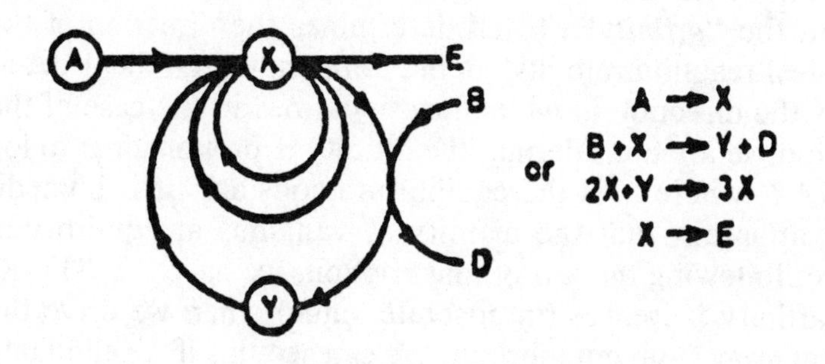

Figure 3. This graph represents the reaction paths for the "Brusselator" reactions, which are further described in the text.

has established that these loops play an essential role in metabolic functions. For example, the relation between nucleic acids and proteins can be described in terms of a crosscatalytic effect: nucleic acids contain the information to produce proteins, which in turn produce nucleic acids.

In addition to the rates of chemical reactions, we must also consider the rates of other irreversible processes, such as heat transfer and the diffusion of matter. The rates of irreversible processes are also called *fluxes* and are denoted by the symbol J. There is no general theory from which we can derive the form of the rates or fluxes. In chemical reactions the rate depends on the molecular mechanism, as can be verified by the examples already indicated. The thermodynamics of irreversible processes introduces a second type of quantity: in addition to the rates, or fluxes, J, it uses "generalized forces," X, that "cause" the fluxes. The simplest example is that of heat conduction. Fourier's law tells us that the heat flux J is proportional to the temperature gradient. This temperature gradient is the "force" causing the heat flux. By definition, flux and forces both vanish at thermal equilibrium. As we shall see, the production of entropy $P = d_iS/dt$ can be calculated from the flux and the forces.

Let us consider the definition of the generalized force corresponding to a chemical reaction. Recall the reaction $A + X$

$\rightarrow Y+B$. We have seen how, at equilibrium, the ratio between concentrations is given by the law of mass action. As Théophile De Donder has shown, a "chemical force" can be introduced, the *"affinity"* $\mathcal{A}$ that determines the direction of the chemical reaction rate just as the temperature gradient determines the direction in which heat will flow. In the case of the reaction we are considering, the affinity is proportional to log KBY/AX, where K is the equilibrium constant. It is immediately apparent that the affinity $\mathcal{A}$ vanishes at equilibrium where, following the law of mass action, we have $AX/BY=K$. The affinity increases (in absolute value) when we drive the system away from equilibrium. We can see this if we eliminate from the system a fraction of the molecules B once they are formed through the reaction $A+X\rightarrow Y+B$. Affinity can be said to measure the distance between the actual state of the system and its equilibrium state. Moreover, as we have mentioned, its sign determines the direction of the chemical reaction. If $\mathcal{A}$ is positive, then there are "too many" molecules B and Y, and the net reaction proceeds in the direction $B+Y\rightarrow A+X$. On the contrary, if $\mathcal{A}$ is negative there are "too few" B and Y, and the net reaction proceeds in the opposite direction.

Affinity as we have defined it is a way of rendering more precise the ancient affinity described by the alchemists, who deciphered the elective relationships between chemical bodies—that is, the "likes" and "dislikes" of molecules. The idea that chemical activity cannot be reduced to mechanical trajectories, to the calm domination of dynamic laws, has been emphasized from the beginning. We could cite Diderot at length. Later, Nietzsche, in a different context, asserted that it was ridiculous to speak of "chemical laws," as though chemical bodies were governed by laws similar to moral laws. In chemistry, he protested, there is no constraint, and each body does as it pleases. It is not a matter of "respect" but of a power struggle, of the ruthless domination of the weaker by the stronger.[2] Chemical equilibrium, with vanishing affinity, corresponds to the resolution of this conflict. Seen from this point of view, the specificity of thermodynamic affinity thus rephrases an age-old problem in modern language,[3] the problem of the distinction between the legal and indifferent world of dynamic law, and the world of spontaneous and productive activity to which chemical reactions belong.

Let us emphasize the basic conceptual distinction between physics and chemistry. In classical physics we can at least conceive of reversible processes such as the motion of a frictionless pendulum. To neglect irreversible processes in dynamics always corresponds to an idealization, but, at least in some cases, it is a meaningful one. The situation in chemistry is quite different. Here the processes that define chemistry—chemical transformations characterized by reaction rates—are irreversible. For this reason chemistry cannot be reduced to the idealization that lies at the basis of classical or quantum mechanics, in which past and future play equivalent roles.

As could be expected, all possible irreversible processes appear in entropy production. Each of them enters through the product of its rate or flux J multiplied by the corresponding force X. The total entropy production per unit time, $P = d_iS/dt$, is the sum of these contributions.

We can divide thermodynamics into three large fields, the study of which corresponds to three successive stages in its development. Entropy production, the fluxes, and the forces are all zero *at equilibrium*. In the *close-to-equilibrium* region, where thermodynamic forces are "weak," the rates J_k are linear functions of the forces. The third field is called the *"nonlinear"* region, since in it the rates are in general more complicated functions of the forces. Let us first emphasize some general features of linear thermodynamics that apply to close-to-equilibrium situations.

Linear Thermodynamics

In 1931, Lars Onsager discovered the first general relations in nonequilibrium thermodynamics for the linear, near-to-equilibrium region. These are the famous "reciprocity relations." In qualitative terms, they state that if a force—say, "one" (corresponding, for example, to a temperature gradient)—may influence a flux "two" (for example, a diffusion process), then force "two" (a concentration gradient) will also influence the flux "one" (the heat flow). This has indeed been verified. For example, in each case where a thermal gradient induces a process of diffusion of matter, we find that a concentration gradient can set up a heat flux through the system.

The *general* nature of Onsager's relations has to be emphasized. It is immaterial, for instance, whether the irreversible processes take place in a gaseous, liquid, or solid medium. The reciprocity expressions are valid independently of any microscopic assumptions.

Reciprocity relations have been the first results in the thermodynamics of irreversible processes to indicate that this was not some ill-defined no-man's-land but a worthwhile subject of study whose fertility could be compared with that of equilibrium thermodynamics. Equilibrium thermodynamics was an achievement of the nineteenth century, nonequilibrium thermodynamics was developed in the twentieth century, and Onsager's relations mark a crucial point in the shift of interest away from equilibrium toward nonequilibrium.

A second general result in this field of *linear,* nonequilibrium thermodynamics bears mention here. We have already spoken of thermodynamic potentials whose extrema correspond to the states of equilibrium toward which thermodynamic evolution tends irreversibly. Such are the entropy S for isolated systems, and the free energy F for closed systems at a given temperature. The thermodynamics of close-to-equilibrium systems also introduces such a potential function. It is quite remarkable that this potential is the *entropy-production* P itself. The theorem of minimum entropy production does, in fact, show that in the range of validity of Onsager's relations—that is, the linear region—a system evolves toward a stationary state characterized by the *minimum* entropy production compatible with the constraints imposed upon the system. These contraints are determined by the boundary conditions. They may, for instance, correspond to two points in the system kept at different temperatures, or to a flux of matter that continuously supports a reaction and eliminates its products.

The stationary state toward which the system evolves is then necessarily a nonequilibrium state at which dissipative processes with nonvanishing rates occur. But since it is a stationary state, all the quantities that describe the system, such as temperature concentrations, become time-independent. Similarly, the entropy of the system now becomes independent of time. Therefore its time variation $dS = 0$ vanishes. But we have seen that the time variation of entropy is made up of two terms—the entropy flow d_eS and the positive entropy production d_iS. Therefore, $dS = 0$ implies that $d_eS = -d_iS < 0$. The heat or matter flux coming from the environment determines a negative flow of entropy d_eS, which is, however, matched by the entropy production d_iS due to irreversible processes inside the system. A negative flux d_eS means that the system transfers entropy to the outside world. Therefore at the stationary state, the system's activity continuously increases the entropy of its environment. This is true for all stationary states. But the theorem of minimum entropy production says more. The particular stationary state toward which the system tends is the one in which this transfer of entropy to the environment is as small as is compatible with the imposed boundary conditions. In this context, the equilibrium state corresponds to the special case that occurs when the boundary conditions allow a vanishing entropy production. In other words, the theory of minimum entropy production expresses a kind of "inertia." When the boundary conditions prevent the system from going to equilibrium it does the next best thing; it goes to a state of minimum entropy production—that is, to a state as close to equilibrium as "possible."

Linear thermodynamics thus describes the stable, predictable behavior of systems tending toward the minimum level of activity compatible with the fluxes that feed them. The fact that linear thermodynamics, like equilibrium thermodynamics, may be described in terms of a potential, the entropy production, implies that, both in evolution toward equilibrium and in evolution toward a stationary state, initial conditions are forgotten. Whatever the initial conditions, the system will finally reach the state determined by the imposed boundary conditions. As a result, the reaction of such a system to any change in its boundary conditions is entirely predictable.

We see that in the linear range the situation remains basically the same as at equilibrium. Although the entropy production does not vanish, neither does it prevent the irreversible change from being identified as an evolution toward a state that is wholly deducible from general laws. This "becoming" inescapably leads to the destruction of any difference, any specificity. Carnot or Darwin? The paradox mentioned in Chapter IV remains. There is still no connection between the appearance of natural organized forms on one side, and on the other the tendency toward "forgetting" of initial conditions, along with the resulting disorganization.

Far from Equilibrium

At the root of nonlinear thermodynamics lies something quite surprising, something that first appeared to be a failure: in spite of much effort, the generalization of the theorem of minimum entropy production for systems in which the fluxes are no longer linear functions of the forces appeared impossible. Far from equilibrium, the system may still evolve to some steady state, but in general this state can no longer be characterized in terms of some suitably chosen potential (such as entropy production for near-equilibrium states).

The absence of any potential function raises a new question: What can we say about the stability of the states toward which the system evolves? Indeed, as long as the attractor state is defined by the minimum of a potential such as the entropy production, its stability is guaranteed. It is true that a fluctuation may shift the system away from this minimum. The second law of thermodynamics, however, imposes the return toward the attractor. The system is thus "immune" with respect to fluctuations. Thus whenever we define a potential, we are describing a *"stable world"* in which systems follow an evolution that leads them to a static situation that is established once and for all.

When the thermodynamic forces acting on a system become such that the linear region is exceeded, however, the stability of the stationary state, or its independence from fluctuations, can no longer be taken for granted. Stability is no longer the consequence of the general laws of physics. We must examine the way a stationary state reacts to the different types of fluctuation produced by the system or its environment. In some cases, the analysis leads to the conclusion that a state is "unstable"—in such a state, certain fluctuations, instead of regressing, may be amplified and invade the entire system, compelling it to evolve toward a new regime that may be qualitatively quite different from the stationary states corresponding to minimum entropy production.

Thermodynamics leads to an initial general conclusion concerning systems that are liable to escape the type of order governing equilibrium. These systems have to be "far from equilibrium." In cases where instability is possible, we have to ascertain the threshold, the distance from equilibrium, at which fluctuations may lead to new behavior, different from the "normal" stable behavior characteristic of equilibrium or near-equilibrium systems.

Why is this conclusion so interesting?

Phenomena of this kind are well known in the field of hydrodynamics and fluid flow. For instance, it has long been known that once a certain flow rate of flux has been reached, turbulence may occur in a fluid. Michel Serres has recently recalled[4] that the early atomists were so concerned about turbulent flow that it seems legitimate to consider turbulence as a basic source of inspiration of Lucretian physics. Sometimes, wrote Lucretius, at uncertain times and places, the eternal, universal fall of the atoms is disturbed by a very slight deviation—the "clinamen." The resulting vortex gives rise to the world, to all natural things. The clinamen, this spontaneous, unpredictable deviation, has often been criticized as one of the main weaknesses of Lucretian physics, as being something introduced *ad hoc*. In fact, the contrary is true—the clinamen attempts to explain events such as laminar flow ceasing to be stable and spontaneously turning into turbulent flow. Today hydrodynamic experts test the stability of fluid flow by introducing a perturbation that expresses the effect of molecular disorder added to the average flow. We are not so far from the clinamen of Lucretius!

For a long time turbulence was identified with disorder or noise. Today we know that this is not the case. Indeed, while turbulent motion appears as irregular or chaotic on the macroscopic scale, it is, on the contrary, highly organized on the microscopic scale. The multiple space and time scales involved in turbulence correspond to the coherent behavior of millions and millions of molecules. Viewed in this way, the transition from laminar flow to turbulence is a process of self-organization. Part of the energy of the system, which in laminar flow was in the thermal motion of the molecules, is being transferred to macroscopic organized motion.

The "Bénard instability" is another striking example of the instability of a stationary state giving rise to a phenomenon of spontaneous self-organization. The instability is due to a vertical temperature gradient set up in a horizontal liquid layer. The lower surface of the latter is heated to a given temperature, which is higher than that of the upper surface. As a result of these boundary conditions, a permanent heat flux is set up, moving from the bottom to the top. When the imposed gradient reaches a threshold value, the fluid's state of rest—the stationary state in which heat is conveyed by conduction alone, without convection—becomes unstable. A convection corresponding to the coherent motion of ensembles of molecules is produced, increasing the rate of heat transfer. Therefore, for given values of the constraints (the gradient of temperature), the entropy production of the system is increased; this contrasts with the theorem of minimum entropy production. The Bénard instability is a spectacular phenomenon. The convection motion produced actually consists of the complex spatial organization of the system. Millions of molecules move coherently, forming hexagonal convection cells of a characteristic size.

In Chapter IV we introduced Boltzmann's order principle, which relates entropy to probability as expressed by the number of complexions P. Can we apply this relation here? To each distribution of the velocities of the molecules corresponds a number of complexions. This number measures the number of ways in which we can realize the velocity distribution by attributing some velocity to each molecule. The argument runs parallel to that in Chapter IV, where we expressed the number of complexions in terms of the distributions of molecules between two boxes. Here also the number of complexions is large when there is disorder—that is, a wide dispersion of velocities. In contrast, coherent motion means that many molecules travel with nearly the same speed (small dispersion of velocities). To such a distribution corresponds a number of complexions *P* so low that there seems almost no chance for the phenomenon of self-organization to occur. Yet it occurs! We see, therefore, that calculating the number of complexions, which entails the hypothesis of an equal a priori probability for each molecular state, is misleading. Its irrelevance is particularly obvious as far as the genesis of the new behavior is concerned. In the case of the Bénard instability it is a fluctuation, a microscopic convection current, which would have been doomed to regression by the application of Boltzmann's order principle, but which on the contrary is amplified until it invades the whole system. Beyond the critical value of the imposed gradient, a new molecular order has thus been produced spontaneously. It corresponds to a giant fluctuation stabilized through energy exchanges with the outside world.

In far-from-equilibrium conditions, the concept of probability that underlies Boltzmann's order principle is no longer valid in that the structures we observe do not correspond to a maximum of complexions. Neither can they be related to a minimum of the free energy $F=E-TS$. The tendency toward leveling out and forgetting initial conditions is no longer a general property. In this context, the age-old problem of the origin of life appears in a different perspective. It is certainly true that life is incompatible with Boltzmann's order principle but not with the kind of behavior that can occur in far-from-equilibrium conditions.

Classical thermodynamics leads to the concept of "equilibrium structures" such as crystals. Bénard cells are structures too, but of a quite different nature. That is why we have introduced the notion of "dissipative structures," to emphasize the close association, at first paradoxical, in such situations between structure and order on the one side, and dissipation or waste on the other. We have seen in Chapter IV that heat transfer was considered a source of waste in classical thermodynamics. In the Bénard cell it becomes a source of order.

The interaction of a system with the outside world, its embedding in nonequilibrium conditions, may become in this way the starting point for the formation of new dynamic states of matter—dissipative structures. Dissipative structures actually correspond to a form of supramolecular organization. Although the parameters describing crystal structures may be derived from the properties of the molecules of which they are composed, and in particular from the range of their forces of attraction and repulsion, Bénard cells, like all dissipative structures, are essentially a reflection of the global situation of nonequilibrium producing them. The parameters describing them are macroscopic; they are not of the order of 10^{-8} cm, like the distance between the molecules of a crystal, but of the order of centimeters. Similarly, the time scales are different—they correspond not to molecular times (such as periods of vibration of individual molecules, which may correspond to about 10^{-15} sec) but to macroscopic times: seconds, minutes, or hours.

Let us return to the case of chemical reactions. There are some fundamental differences from the Bénard problem. In the Bénard cell the instability has a simple mechanical origin. When we heat the liquid layer from below, the lower part of the fluid becomes less dense, and the center of gravity rises. It is therefore not surprising that beyond a critical point the system tilts and convection sets in.

But in chemical systems there are no mechanical features of this type. Can we expect any self-organization? Our mental image of chemical reactions corresponds to molecules speeding through space, colliding at random in a chaotic way. Such an image leaves no place for self-organization, and this may be one of the reasons why chemical instabilities have only recently become a subject of interest. There is also another difference. *All* flows become turbulent at a "sufficiently" large distance from equilibrium (the threshold is measured by dimensionless numbers such as Reynolds' number). This is not true for chemical reactions. Being far from equilibrium is a necessary requirement but not a sufficient one. For many chemical systems, whatever the constraints imposed and the rate of the chemical changes produced, *the stationary state remains stable* and arbitrary fluctuations are damped, as is the case in the close-to-equilibrium range. This is true in particular of systems in which we have a chain of transformations of the type $A \rightarrow B \rightarrow C \rightarrow D$ . . . and that may be described by *linear* differential equations.

The fate of the fluctuations perturbing a chemical system, as well as the kinds of new situations to which it may evolve, thus depend on the detailed mechanism of the chemical reactions. In contrast with close-to-equilibrium situations, the behavior of a far-from-equilibrium system becomes highly specific. There is no longer any universally valid law from which the overall behavior of the system can be deduced. Each system is a separate case; each set of chemical reactions must be investigated and may well produce a qualitatively different behavior.

Nevertheless, one general result has been obtained, namely a necessary condition for chemical instability: in a chain of chemical reactions occurring in the system, *the only reaction stages* that, under certain conditions and circumstances, may jeopardize the stability of the stationary state are precisely the "catalytic loops"—stages in which the product of a chemical reaction is involved in its own synthesis. This is an interesting conclusion, since it brings us closer to some of the fundamental achievements of modern molecular biology (see Figure 4).

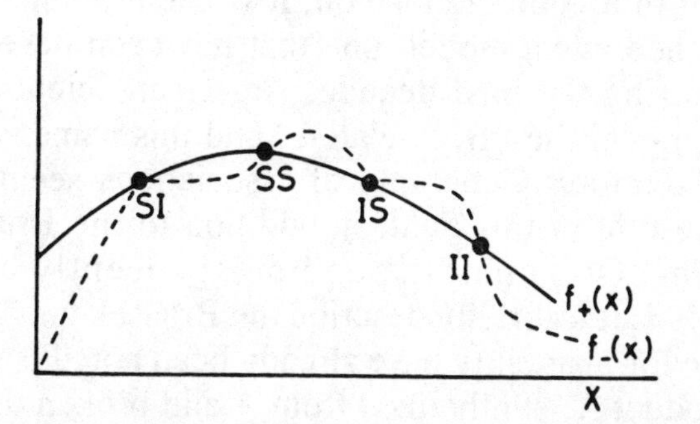

Figure 4. Catalytic loops correspond to nonlinear terms. In the case of a one-independent-variable problem, this means the occurrence of at least one term where the independent variable appears with a power higher than 1; in this simple case, it is easy to see the relation between such nonlinear terms and the potential instability of stationary states.

Let us take for the independent variable X the time evolution $dX/dt = f(X)$. It is always possible to decompose $f(X)$ in two functions representing a gain and a loss $f_+(X)$ and $f_-(X)$, each of which is positive or 0, such that $f(X) = f_+(X) - f_-(X)$. In this way, stationary states ($dX/dt = 0$) correspond to values where $f_+(X) = f_-(X)$.

Those states are graphically given by the intersections of the two graphs plotting f_+ and f_-. If f_+ and f_- are linear, there can only be one intersection. In other cases, the type of the intersection permits us to infer the stability of the stationary state.

Four cases are possible:

SI: stable with respect to negative fluctuations, unstable with respect to positive ones: If the system deviates slightly to the left of SI, the positive difference between f_+ and f_- will reduce this deviation back to SI; deviations to the right will be amplified.

SS: stable with respect to positive and negative fluctuations.

IS: stable only with respect to positive fluctuations.

II: unstable with respect to positive and negative fluctuations.

Beyond the Threshold of Chemical Instability

Today the study of chemical instabilities is common. Both theoretical and experimental work are being pursued in a large number of institutions and laboratories. Indeed, as will become clear, these investigations are of interest to a wide range of scientists—not only to mathematicians, physicists, chemists, and biologists, but also to economists and sociologists.

In far-from-equilibrium conditions various new phenomena appear beyond the threshold of chemical instability. To describe them in a concrete fashion, it is useful to start with a simplified theoretical model, one that has been developed at Brussels during the past decade. American scientists have called this model the "Brusselator," and this name is used in scientific literature (Geographical associations seem to have become the rule in this field; in addition to the Brusselator, there is an "Oregonator," and most recently a "Paloaltonator"!). Let us briefly describe the Brusselator. The steps responsible for instability have already been noted (see Figure 3). The product *X*, synthetized from *A* and broken down into the form of *E*, is linked by a relationship of crosscatalysis to produce *Y*. *X* is produced from *Y* during a trimolecular step but, conversely, *Y* is synthetized by a reaction between *X* and a reactant *B*.

In this model, the concentrations of the products and the reactants *A*, *B*, *D*, and *E* are given parameters (the "control substances"). The behavior of the system is explored for increasing values of *B*, with *A* remaining constant. The stationary state toward which such a system is likely to evolve—the state for which $dX/dt = dY/dt = 0$—corresponds to concentrations $X_0 = A$ *and* $Y_0 = B/A$. This can be easily verified by writing the kinetic equations and looking for the stationary state. However, this stationary state ceases to be stable as soon as the concentration of *B* exceeds a critical threshold (everything else being kept equal). After the critical threshold has been reached, the stationary state becomes an unstable "focus" and the system leaves this focus to reach a "limit cycle."

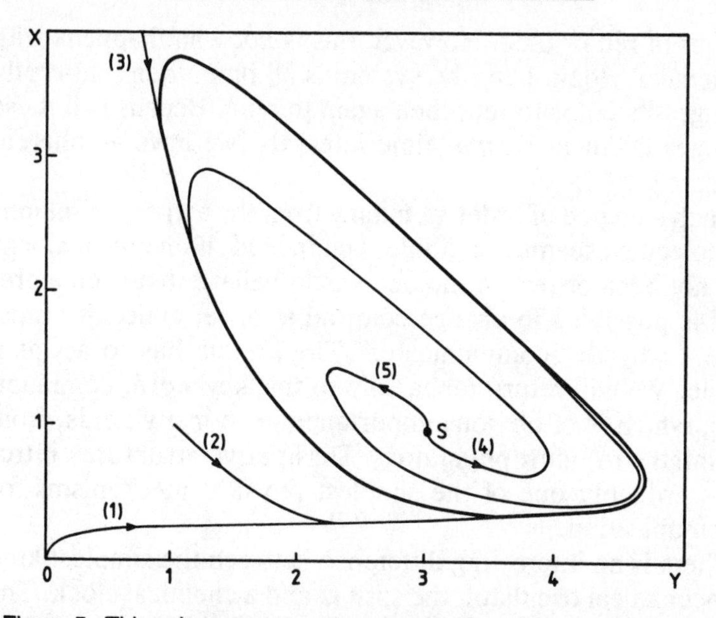

Figure 5. This scheme represents concentration of component *X* vs. concentration of component *Y*. The cycle's focus (point *S*) is the stationary state, which is unstable for $B>(1+A^2)$. All the trajectories (of which five are plotted), whatever their intitial state, lead to the same cycle.

Instead of remaining stationary, the concentrations of *X* and *Y* begin to oscillate with a well-defined periodicity. The oscillation period depends both on the kinetic constants characterizing the reaction rates and the boundary conditions imposed on the system as a whole (temperature, concentration of *A*, *B*, etc.).

Beyond the critical threshold the system spontaneously leaves the stationary state $X_0=A$, $Y_0=B/A$ as the result of fluctuations. Whatever the initial conditions, it approaches the limit cycle, the periodic behavior of which is stable. We therefore have a periodic chemical process—a chemical clock. Let us pause a moment to emphasize how unexpected such a phenomenon is. Suppose we have two kinds of molecules, "red" and "blue." Because of the chaotic motion of the molecules, we would expect that at a given moment we would have more red molecules, say, in the left part of a vessel. Then a bit later more blue molecules would appear, and so on. The vessel would appear to us as "violet," with occasional irregular flashes of red or blue. However, this is *not* what happens with a chemical clock; here the system is all blue, then it abruptly changes its color to red, then again to blue. Because all these changes occur at *regular* time intervals, we have a coherent process.

Such a degree of order stemming from the activity of billions of molecules seems incredible, and indeed, if chemical clocks had not been observed, no one would believe that such a process is possible. To change color all at once, molecules must have a way to "communicate." The system has to act as a whole. We will return repeatedly to this key word, communicate, which is of obvious importance in so many fields, from chemistry to neurophysiology. Dissipative structures introduce probably one of the simplest physical mechanisms for communication.

There is an interesting difference between the simplest kind of mechanical oscillator, the spring, and a chemical clock. The chemical clock has a well-defined periodicity corresponding to the limit cycle its trajectory is following. On the contrary, a spring has a frequency that is amplitude-dependent. From this point of view a chemical clock is more reliable as a timekeeper than a spring.

But chemical clocks are not the only type of self-organization. Until now diffusion has been neglected. All substances were assumed to be evenly distributed over the reaction space. This is an idealization; small fluctuations will always lead to differences in concentrations and thus to diffusion. We therefore have to add diffusion to the chemical reaction equations. The diffusion-reaction equations of the Brusselator display an astonishing range of behaviors available to this system. Indeed, whereas at equilibrium and near-equilibrium the system remains spatially homogeneous, the diffusion of the chemical throughout the system induces, in the far-from-equilibrium region, the possibility of new types of instability, including the amplification of fluctuations breaking the initial spatial symmetry. Oscillations in time, chemical clocks, thus cease to be the only kind of dissipative structure available to the system. Far from it; for example, oscillations may appear that are now both time- and space-dependent. They correspond to chemical waves of X and Y concentrations that periodically pass through the system.

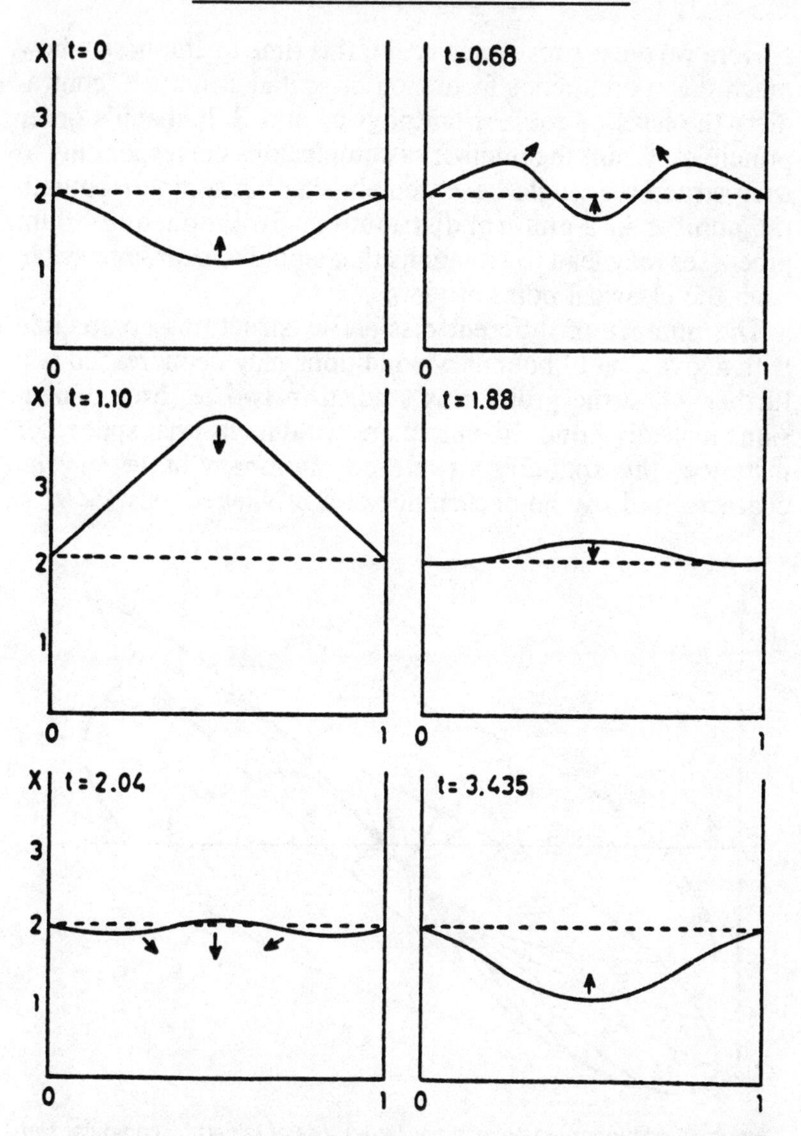

Figure 6. Chemical waves simulated on computer: successive steps of evolution of spatial profile of concentration of constituent X in the "Brusselator" trimolecular model. At time $t = 3.435$ we recover the same distribution as at time $t = 0$. Concentration of A and B: 2, 5.45 ($B > [1 + A^2]$). Diffusion coefficients for X and Y: 8 10^{-3}, 4 10^{-3}.

In addition, especially when the values of the diffusion constants of X and Y are quite different from each other, the system may display a stationary, time-independent behavior, and stable spatial structures may appear.

Here we must pause once again, this time to emphasize how much the spontaneous formation of spatial structures contradicts the laws of equilibrium physics and Boltzmann's order principle. Again, the number of complexions corresponding to such structures would be extremely small in comparison with the number in a uniform distribution. Still, nonequilibrium processes may lead to situations that would appear impossible from the classical point of view.

The number of different dissipative structures compatible with a given set of boundary conditions may be increased still further when the problem is studied in two or three dimensions instead of one. In a circular, two-dimensional space, for instance, the spatially structured stationary state may be characterized by the occurrence of a privileged axis.

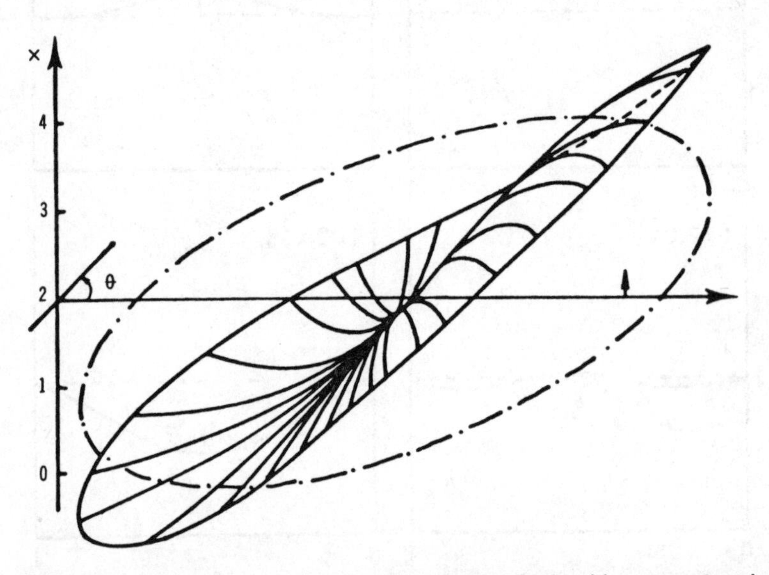

Figure 7. Stationary state with privileged axis obtained by computer simulation. Concentration X is a function of geometrical coordinates ρ,θ in the horizontal plane. The location of the perturbation applied to the uniform unstable solution (X_0, Y_0) is indicated by an arrow.

This corresponds to a new, extremely interesting symmetry-breaking process, especially when we recall that one of the first stages in morphogenesis of the embryo is the formation of a gradient in the system. We will return to these problems later in this chapter and again in Chapter VI.

Up to now it has been assumed that the "control substances" ($A, B, D,$ and E) are uniformly distributed throughout the reaction system. If this simplification is abandoned, additional phenomena can occur. For example, the system takes on a "natural size," which is a function of the parameters describing it. In this way the system determines its own intrinsic size—that is, it determines the region that is spatially structured or crossed by periodic concentration waves.

These results still give a very incomplete picture of the variety of phenomena that may occur far from equilibrium. Let us first mention the possibility of multiple states far from equilibrium. For given boundary conditions there may appear more than one stationary state—for instance one rich in the chemical X, the other poor. The shift from one state to another plays an important role in control mechanisms as they have been described in biological systems.

Since the classical work of Lyapounov and Poincaré, characteristic points such as focus or lines such as limit cycles

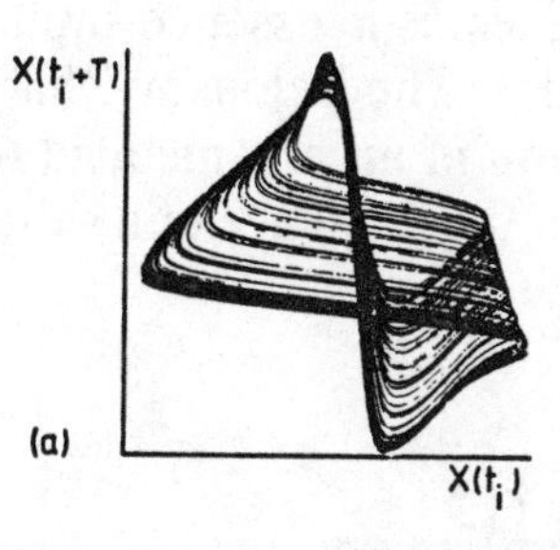

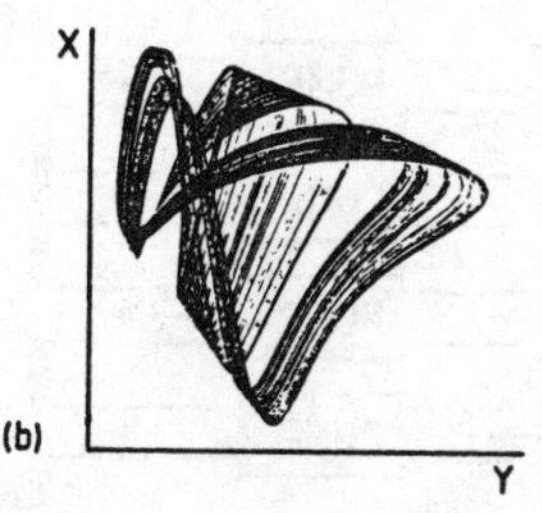

Figure 8. (a) Bromide-ion concentration in the Belousov-Zhabotinsky reaction at times t_i and $t_i + T$ (cf. R. H. Simoyi, A. Wolf, and H. L. Swinney, *Physical Review Letters*, Vol. 49 (1982), p. 245; see J. Hirsch, "Condensed Matter Physics," and on computers, *Physics Today* (May 1983), pp. 44–52).

(b) Attractor lines calculated by Hao Bai-lin for a Brusselator with external periodic supply of component X (personal communication).

were known to mathematicians as the "attractors" of stable systems. What is new is their application to chemical systems. It is worth noting that the first paper dealing with instabilities in reaction-diffusion systems was published by Turing in 1952. In recent years new types of attractors have been identified. They appear only when the number of independent variables increases (there are two independent variables in the Brusselator, the variables *X* and *Y*). In particular, we can get "strange attractors" that do not correspond to periodic behavior.

Figure 8(b), which summarizes some calculations by Hao Bailin, gives an idea of such very complicated attractor lines calculated for a model generalizing the Brusselator through the addition of an external periodic supply of *X*. What is remarkable is that most of the possibilities we have described have been observed in inorganic chemistry as well as in a number of biological situations.

In inorganic chemistry the best-known example is the Belousov-Zhabotinsky reaction discovered in the early 1960s. The corresponding reaction scheme, the Oregonator, introduced by Noyes and his colleagues, is in essence similar to the Brusselator though more complex. The Belousov-Zhabotinsky reaction consists of the oxidation of an organic acid (malonic acid) by a potassium bromate in the presence of a suitable catalyst, cerium, manganese, or ferroin.

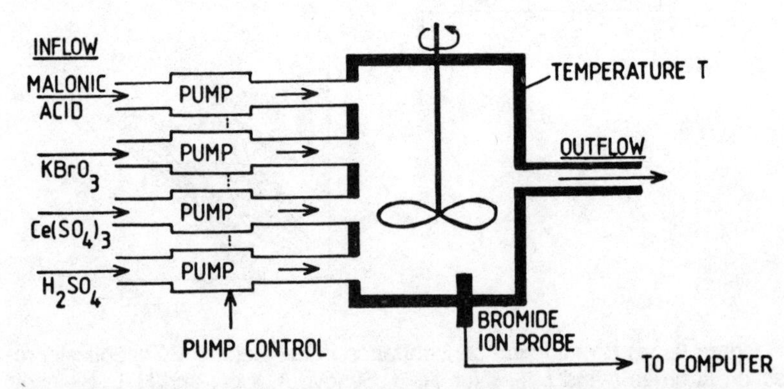

Figure 9. Schematic representation of a chemical reactor used to study the oscillations in the Belousov-Zhabotinsky reaction (there is a stirring device in the reactor to keep the system homogeneous). The reaction has over thirty products and intermediates. The evolution of different reaction paths depends (among others factors) on the entries controlled by the pumps.

Various experimental conditions may be set up giving different forms of autoorganization within the same system—a chemical clock, a stable spatial differentiation, or the formation of waves of chemical activity over macroscopic distances.[5]

Let us now turn to a matter of the greatest interest: the relevance of these results for the understanding of living systems.

The Encounter with Molecular Biology

Earlier in this chapter we showed that in far-from-equilibrium conditions various types of self-organization processes may occur. They may lead to the appearance of chemical oscillations or to spatial structures. We have seen that the basic condition for the appearance of such phenomena is the existence of catalytic effects.

Although the effects of "nonlinear" reactions (the presence of the reaction product) have a feedback action on their "cause" and are comparatively rare in the inorganic world, molecular biology has discovered that they are virtually the rule as far as living systems are concerned. Autocatalysis (the presence of X accelerates its own synthesis), autoinhibition (the presence of X blocks a catalysis needed to synthesize it), and crosscatalysis (two products belonging to two different reaction chains activate each other's synthesis) provide the classical regulation mechanism guaranteeing the coherence of the metabolic function.

Let us emphasize an interesting difference. In the examples known in inorganic chemistry, the molecules involved are simple but the reaction mechanisms are complex—in the Belousov-Zhabotinsky reaction, about thirty compounds have been identified. On the contrary, in the many biological examples we have, the reaction scheme is simple but the molecules (proteins, nucleic acids, etc.) are highly complex and specific. This can hardly be an accident. Here we encounter an initial element marking the difference between physics and biology. Biological systems *have a past*. Their constitutive molecules are the result of an evolution; they have been selected to take part in the autocatalytic mechanisms to generate very specific forms of organization processes.

A description of the network of metabolic activations and

inhibitions is an essential step in understanding the functional logic of biological systems. This includes the triggering of syntheses at the moment they are needed and the blocking of those chemical reactions whose unused products would accumulate in the cell.

The basic mechanism through which molecular biology explains the transmission and exploitation of genetic information is itself a feedback loop, a "nonlinear" mechanism. Deoxyribonucleic acid (DNA), which contains in sequential form all the information required for the synthesis of the various basic proteins needed in cell building and functioning, participates in a sequence of reactions during which this information is *translated* into the form of different protein sequences. Among the proteins synthesized, some enzymes exert a feedback action that activates or controls not only the different transformation stages but also the autocatalytic mechanism of DNA replication, by which genetic information is copied at the same rate as the cells multiply.

Here we have a remarkable case of the convergence of two sciences. The understanding attained here required complementary developments in physics and biology, one toward the complex and the other toward the elementary.

Indeed, from the point of view of physics, we now investigate "complex" situations far removed from the ideal situations that can be described in terms of equilibrium thermodynamics. On the other hand, molecular biology succeeded in relating living structures to a relatively small number of basic biomolecules. Investigating the diversity of chemical mechanisms, it discovered the intricacy of the metabolic reaction chains, the subtle, complex logic of the control, inhibition, and activation of the catalytic function of the enzymes associated with the critical step of each of the metabolic chains. In this way molecular biology provides the microscopic basis for the instabilities that may occur in far-from-equilibrium conditions.

In a sense, living systems appear as a well-organized factory: on the one hand, they are the site of multiple chemical transformations; on the other, they present a remarkable "space-time" organization with highly nonuniform distribution of biochemical material. We can now link function and structure. Let us briefly consider two examples that have been studied extensively in the past few years.

First we shall consider glycolysis, the chain of metabolic reactions during which glucose is broken down and an energy-rich substance ATP (adenosine triphosphate) is synthetized, providing an essential source of energy common to all living cells. For each glucose molecule that is broken down, two molecules of ADP (adenosine disphosphate) are transformed into two molecules of ATP. Glycolysis provides a fine example of how complemetary the analytical approach of biology and the investigation of stability in far-from-equilibrium conditions are.[6]

Biochemical experiments have discovered the existence of temporal oscillations in concentrations related to the glycolytic cycle.[7] It has been shown that these oscillations are determined by a key step in the reaction sequence, a step activated by ADP and inhibited by ATP. This is a typical nonlinear phenomenon well suited to regulate metabolic functioning. Indeed, each time the cell draws on its energy reserves, it is exploiting the phosphate bonds, and ATP is converted into ADP. ADP accumulation inside the cell thus signifies intensive energy consumption and the need to replenish stocks. ATP accumulation, on the other hand, means that glucose can be broken down at a slower rate.

Theoretical investigation of this process has shown that this mechanism is indeed liable to produce an oscillation phenomenon, a chemical clock. The theoretically calculated values of the chemical concentrations necessary to produce oscillation and the period of the cycle agree with the experimental data. Glycolytic oscillation produces a modulation of all the cell's energy processes which are dependent on ATP concentration and therefore indirectly on numerous other metabolic chains.

We may go farther and show that in the glycolytic pathway the reactions controlled by some of the key enzymes are in far-from-equilibrium conditions. Such calculations have been reported by Benno Hess[8] and have since been extended to other systems. Under usual conditions the glycolytic cycle corresponds to a chemical clock, but changing these conditions can induce spatial pattern formations in complete agreement with the predictions of existing theoretical models.

A living system appears very complex from the thermodynamic point of view. Certain reactions are close to equilibrium, and others are not. Not everything in a living system is "alive." The energy flow that crosses it somewhat resembles the flow of a river that generally moves smoothly but that from time to time tumbles down a waterfall, which liberates part of the energy it contains.

Let us consider another biological process that also has been studied from the point of view of stability: the aggregation of slime molds, the Acrasiales amoebas *(Dictyostelium discoideum)*. This process[9A] is an interesting case on the borderline between unicellular and pluricellular biology. When

The aggregation of cellular slime molds furnishes a particularly remarkable example of a self-organization phenomenon in a biological system in which a chemical clock plays an essential role. See Figure A.

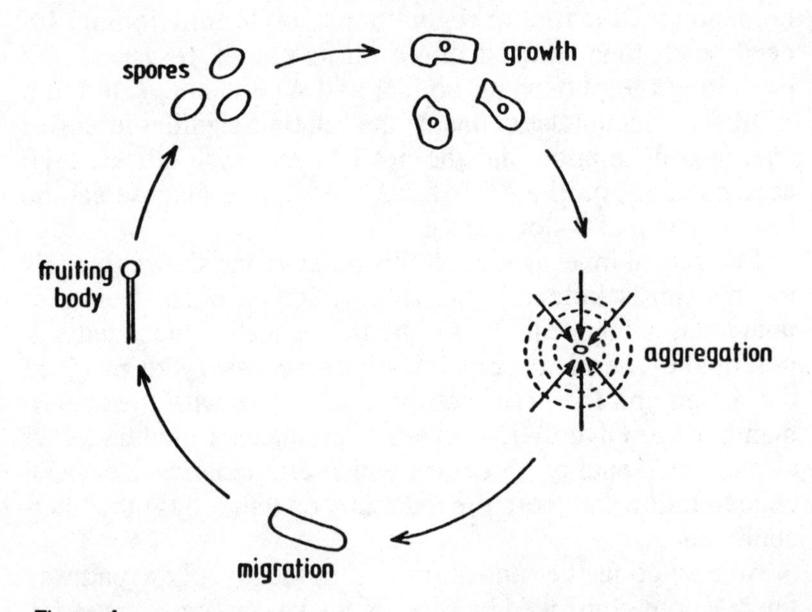

Figure A

When coming out of spores the amoebae grow and multiply as unicellular organisms. This situation extends until food, principally furnished by bacteria, becomes scarce. Then the amoebae cease to reproduce and enter into an interphase that lasts some eight hours. At the end of this period the amoebae begin to aggregate around cells that behave as aggregation centers. The aggregation occurs in response to chemotactic signals emitted by the centers. The aggregate thus formed migrates until the conditions for the formation of a fruiting body are satisfied. Then the mass of cells differentiates to form a stalk surmounted by a mass of spores.

In *Dictyostelium discoideum,* the aggregation proceeds in a periodic manner. Movies of aggregation process show the existence of concentric waves of amoebae moving toward the center with a periodicity of several minutes. The nature of the chemotactic factor is known: it is cyclic AMP (cAMP), a substance involved in numerous biochemical processes such as hormonal regulations. The aggregation centers release the signals of cAMP in a periodic fashion. The other cells respond by moving toward the centers and by relaying the signals to the periphery of the aggregation territory. The existence of a mechanism of relay of the chemotactic signals allows each center to control the aggregation of some 10^5 amoebae.

The analysis of a model of the process of aggregation reveals the existence of two types of bifurcations. First the aggregation itself represents a breaking of spatial symmetry. The second bifurcation breaks the temporal symmetry.

Initially the amoebae are homogeneously distributed. When some of them begin to secrete the chemotactic signals, there appear local fluctuations in the concentration of cAMP. For a critical value of some parameter of the system (diffusion coefficient of cAMP, motility of the amoebae, etc.), fluctuations are amplified: the homogeneous distribution becomes unstable and the amoebae evolve toward an inhomogeneous distribution in space. This new distribution corresponds to the accumulation of amoebae around aggregation centers.

To understand the origin of the periodicity in the aggregation of *D. discoideum,* it is necessary to study the mechanism of synthesis of the chemotactic signal. On the basis of experimental observations one can describe this mechanism by the scheme of Figure B.

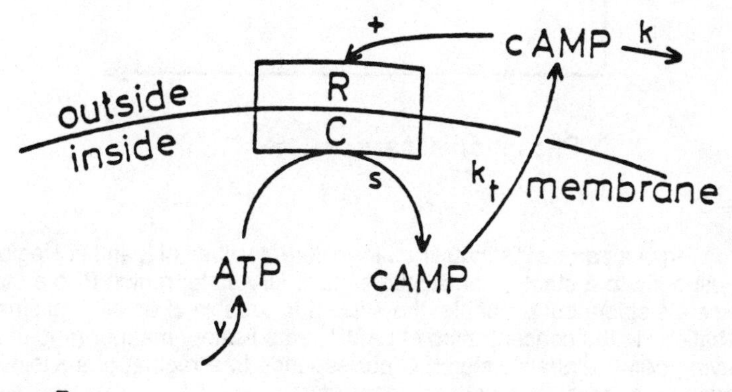

Figure B

On the surface of the cell, receptors (R) bind the molecules of cAMP. The receptor faces the extracellular medium and is functionally linked to an enzyme, adenylate cyclase (C), which transforms intracellular ATP into cAMP. The cAMP thus synthesized is transported across the membrane into the extracellular medium, where it is degraded by phosphodiesterase, an enzyme that is secreted by the amoebae. The experiments show that binding of extracellular cAMP to the membrane receptor activates adenylate cyclase (positive feedback indicated by +).

On the basis of this autocatalytic regulation, the analysis of a model for cAMP synthesis has permitted unification of different types of behavior observed during aggregation.[9B]

Two key parameters of the model are the concentrations of adenylate cyclase (s) and of phosphodiesterase (k). Figure C (redrawn from A. GOLDBETER and L. SEGEL, *Differentiation,* Vol. 17 [1980], pp. 127–35), shows the behavior of the modelized system in the space formed by s and k.

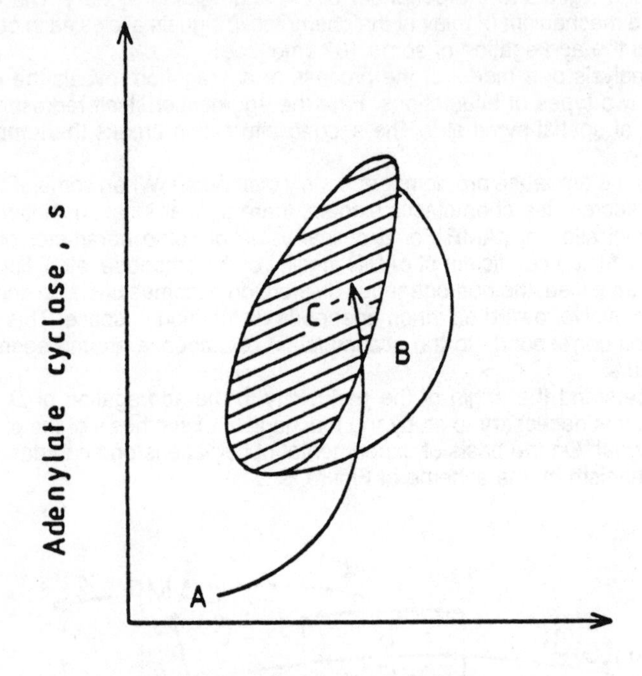

Figure C

Three regions can be distinguished for different values of k and s. Region A corresponds to a stable, nonexcitable stationary state; region B to a stationary state stable but excitable: the system is capable of amplifying small perturbations in the concentration of cAMP in a pulsatory manner (and thus of relaying cAMP signals); region C corresponds to a regime of sustained oscillations around an unstable stationary state.

The arrow indicates a possible "developmental path" corresponding to a rise in phosphodiesterase (k) and adenylate cyclase (s), a rise that is observed to occur after the beginning of starvation. The crossing of regions A, B and C corresponds to the observed change of behavior: cells are at first incapable of responding to extracellular cAMP signals; thereafter they relay these signals and, finally, they become capable of synthetizing them periodically in an autonomous way. The aggregation centers would thus be the cells for which the parameters s and k have reached the more rapidly a point located inside region C after starvation has begun.

the environment in which these amoebas live and multiply becomes poor in nutrients, they undergo a spectacular transformation. (See Figure A.) Starting as a population of isolated cells, they join to form a mass composed of several tens of thousands of cells. This "pseudoplasmodium" then undergoes differentiation, all the while changing shape. A "foot" forms, consisting of about one third of the cells and containing abundant cellulose. This foot supports a round mass of spores, which will detach themselves and spread, multiplying as soon as they come in contact with a suitable nutrient medium and thus forming a new colony of amoebas. This is a spectacular example of adaptation to the environment. The population lives in one region until it has exhausted the available resources. It then goes through a metamorphosis by means of which it acquires the mobility to invade other environments.

An investigation of the first stage of the aggregation process reveals that it begins with the onset of displacement waves in the amoeba population, with a pulsating motion of convergence of the amoebaes toward a "center of attraction," which appears to be produced spontaneously. Experimental investigation and modelization have shown that this migration is a response by the cells to the existence in the environment of a concentration gradient in a key substance, cyclic AMP, which is periodically produced by an amoeba which is the attractor center and later by other cells through a relay mechanism. Here we again see the remarkable role of chemical clocks. They provide, as we have already stressed, new means of communication. In the present case, the self-organization mechanism leads to communication between cells.

There is another aspect we wish to emphasize. Slime mold aggregation is a typical example of what may be termed "order through fluctuations": the setting up of the attractor center giving off the AMP indicates that the metabolic regime corresponding to a normal nutritive environment has become unstable—that is, the nutritive environment has become exhausted. The fact that under such conditions of food shortage any given amoeba can be the first to emit cyclic AMP and thus become an attractor center corresponds to the random behavior of fluctuations. This fluctuation is then amplified and organizes the medium.

Bifurcations and Symmetry-Breaking

Let us take a closer look at the emergence of self-organization and the processes that occur when we go beyond this threshold. At equilibrium or near-equilibrium, there is only one steady state that will depend on the values of some control parameters. We shall call λ the control parameter, which, for example, may be the concentration of substance *B* in the Brusselator described in section 4. We now follow the change in the state of the system as the value of *B* increases. In this way the system is pushed farther and farther away from equilibrium. At some point we reach the threshold of the stability of the "thermodynamic branch." Then we reach what is generally called a "bifurcation point." (These are the points whose role Maxwell emphasized in his thoughts on the relation between determinism and free choice [see Chapter II, section 3].)

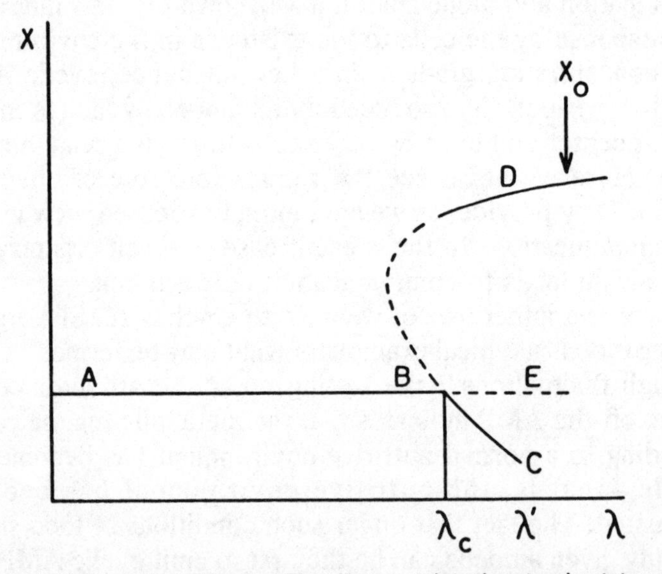

Figure 10. Bifurcation diagram. The diagram plots the steady-state values of X as function of a bifurcation parameter λ. Continuous lines are stable stationary states; broken lines are unstable stationary states. The only way to get to branch D is to start with some concentration X_0 higher than the value of X corresponding to branch E.

Let us consider some typical bifurcation diagrams. At bifurcation point *B*, the thermodynamic branch becomes unstable in respect to fluctuations. For the value λ_c of the control parameter λ, the system may be in three different steady states: *C, E, D*. Two of these states are stable, one unstable. It is very important to emphasize that the behavior of such systems depends on their history. Suppose we slowly increase the value of the control parameter λ; we are likely to follow the path *A, B, C* in Figure 10. On the contrary, if we start with a large value of the concentration *X* and maintain the value of the control parameter constant, we are likely to come to point *D*. The state we reach depends on the previous history of the system. Until now history has been commonly used in the interpretation of biological and social phenomena, but that it may play an important role in simple chemical processes is quite unexpected.

Consider the bifurcation diagram represented in Figure 11. This differs from the previous diagram in that at the bifurcation point *two* new stable solutions emerge. Thus a new question: Where will the system go when we reach the bifurcation point? We have here a "choice" between two possibilities;

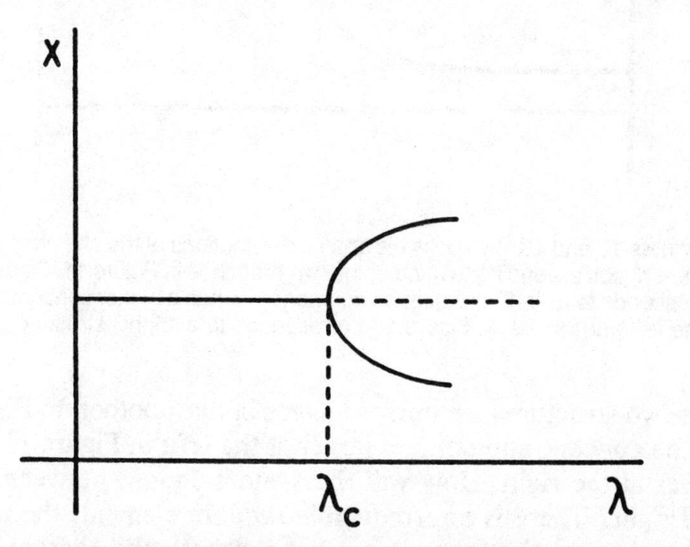

Figure 11. Symmetrical bifurcation diagram. *X* is plotted as a function of λ. For $\lambda<\lambda_c$ there is only one stationary state, which is stable. For $\lambda>\lambda_c$ there are two stable stationary states for each value of λ (the formerly stable state becomes unstable).

they may represent either of the two nonuniform distributions of chemical *X* in space, as represented in Figures 12 and 13.

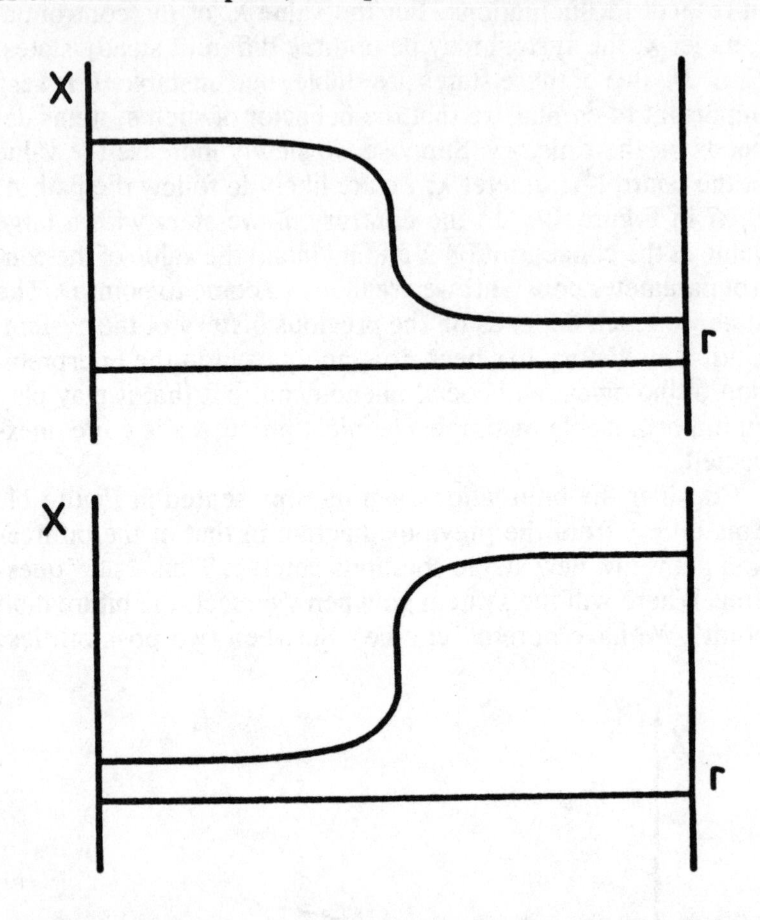

Figures 12 and 13. Two possible spatial distributions of the chemical component *X*, corresponding to each of the two branches in Figure 11. Figure 12 corresponds to a "left" structure as component *X* has a higher concentration in the left part; similarly, Figure 13 corresponds to a "right" structure.

The two structures are mirror images of one another. In Figure 12 the concentration of *X* is larger at the left; in Figure 13 it is larger at the right. How will the system choose between left and right? There is an irreducible random element; the macroscopic equation cannot predict the path the system will take. Turning to a microscopic description will not help. There is also no distinction between left and right. We are faced with chance events very similar to the fall of dice.

We would expect that if we repeat the experiment many times and lead the system beyond the bifurcation point, half of the system will go into the left configuration, half into the right. Here another interesting question arises: In the world around us, some basic simple symmetries seem to be broken.[10] Everybody has observed that shells often have a preferential chirality. Pasteur went so far as to see in dissymmetry, in the breaking of symmetry, the very characteristic of life. We know today that DNA, the most basic nucleic acid, takes the form of a right-handed helix. How did this dissymmetry arise? One common answer is that it comes from a unique event that has by chance favored one of the two possible outcomes; then an autocatalytic process sets in, and the right-handed structure produces other right-handed structures. Others imagine a "war" between left- and right-handed structures in which one of them has annihilated the other. These are problems for which we have not yet found a satisfactory answer. To speak of unique events is not satisfactory; we need a more "systematic" explanation.

We have recently discovered a striking example of the fundamental new properties that matter acquires in far-from-equilibrium conditions: external fields, such as the gravitational field, can be "perceived" by the system, creating the possibility of pattern selection.

How would an external field—a gravitational field—change an equilibrium situation? The answer is provided by Boltzmann's order principle: the basic quantity involved is the ratio of potential energy/thermal energy. This is a small quantity for the gravitational field of earth; we would have to climb a mountain to achieve an appreciable change in pressure or in the composition of the atmosphere. But recall the Bénard cell; from a mechanical perspective, its instability is the raising of its center of gravity as the result of thermal dilatation. In other words, gravitation plays an essential role here and leads to a new structure in spite of the fact that the Bénard cell may have a thickness of only a few millimeters. The effect of gravitation on such a thin layer would be negligible at equilibrium, but because of the nonequilibrium induced by the difference in temperature, macroscopic effects due to gravitation become visible even in this thin layer. Nonequilibrium magnifies the effect of gravitation.[11]

Gravitation obviously will modify the diffusion flow in a reaction diffusion equation. Detailed calculations show that this can be quite dramatic near a bifurcation point of an unperturbed system. In particular, we can conclude that very small gravitational fields can lead to pattern selection.

Let us again consider a system with a bifurcation diagram such as represented in Figure 11. Suppose that for no gravitation, $g=0$, we have, as in Figures 12 and 13, an asymmetric "up/down" pattern as well as its mirror image, "down/up." Both are equally probable, but when g is taken into account, the bifurcation equations are modified because the diffusion flow contains a term proportional to g. As a result, we now obtain the bifurcation diagram represented in Figure 14. The original bifurcation has disappeared—this is true whatever the value of the field. One structure (a) now emerges *continuously* as the bifurcation parameter grows, while the other (b) can be attained only through a finite perturbation.

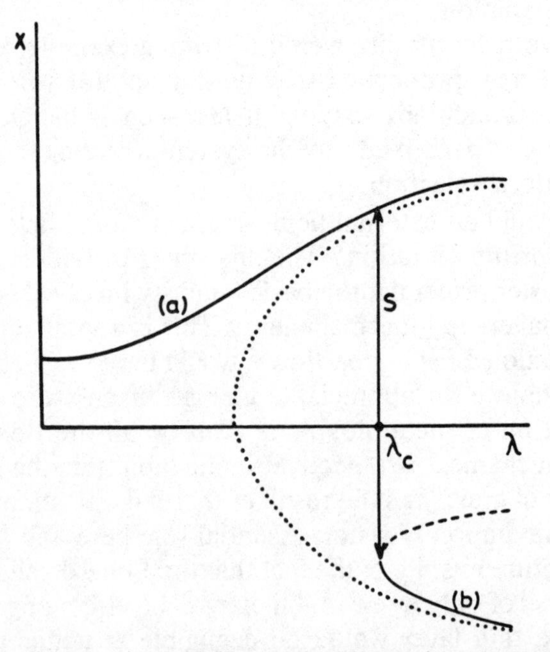

Figure 14. Phenomenon of assisted bifurcation in the presence of an external field. *X* is plotted as a function of parameter λ. The symmetrical bifurcation that would occur in the absence of the field is indicated by the dotted line. The bifurcation value is λ_c; the stable branch (b) is at finite distance from branch (a).

Therefore, if we follow the path (a), we expect the system to follow the continuous path. This expectation is correct as long as the distance s between the two branches remains large in respect to thermal fluctuations in the concentration of *X*. There occurs what we would like to call an "assisted" bifurcation. As before, at about the value λ_c a self-organization process may occur. But now one of the two possible patterns is preferred and will be selected.

The important point is that, depending on the chemical process responsible for the bifurcation, this mechanism expresses an extraordinary sensitivity. Matter, as we mentioned earlier in this chapter, perceives differences that would be insignificant at equilibrium. Such possibilities lead us to think of the simplest organisms, such as bacteria, which we know are able to react to electric or magnetic fields. More generally they show that far-from-equilibrium chemistry leads to possible "adaptation" of chemical processes to outside conditions. This contrasts strongly with equilibrium situations, in which large perturbations or modifications of the boundary conditions are necessary to determine a shift for one structure to another.

The sensitivity of far-from-equilibrium states to external fluctuations is another example of a system's spontaneous "adaptative organization" to its environment. Let us give an example[12] of self-organization as a function of fluctuating external conditions. The simplest conceivable chemical reaction is the isomerization reaction where $A \leftrightarrows B$. In our model the product *A* can also enter into another reaction: $A + light \rightarrow A^* \rightarrow A + heat$. *A* absorbs light and gives it back as heat while leaving its excited state A^*. Consider these two processes as taking place in a closed system: only light and heat can be exchanged with the outside. Nonlinearity exists in the system because the transformation from *B* to *A* absorbs heat: the higher the temperature, the faster the formation of *A*. But also the higher the concentration of *A*, the higher the absorption of light by *A* and its transformation into heat, and the higher the temperature. *A* catalyzes its own formation.

We expect to find that the concentration of *A* corresponding to the stationary state increases with the light intensity. This is indeed the case. But starting from a critical point, there appears one of the standard far-from-equilibrium phenomena: the coexistence of multiple stationary states. For the same values of light intensity and temperature, the system can be found in two different stable stationary states with different concentrations of *A*. A third, unstable state marks the threshold between the first two. Such a coexistence of stationary states gives birth to the well-known phenomenon of hysteresis. But this is not the whole story. If the light intensity, instead of being constant, is taken as randomly fluctuating, the situation is altered profoundly. The zone of coexistence between the two stationary states increases, and for certain values of the parameters coexistence among *three* stationary stable states becomes possible.

In such a case, a random fluctuation in the external flux, often termed "noise," far from being a nuisance, produces new types of behavior, which would imply, under deterministic fluxes, much more complex reaction schemes. It is important to remember that random noise in the fluxes may be consid-

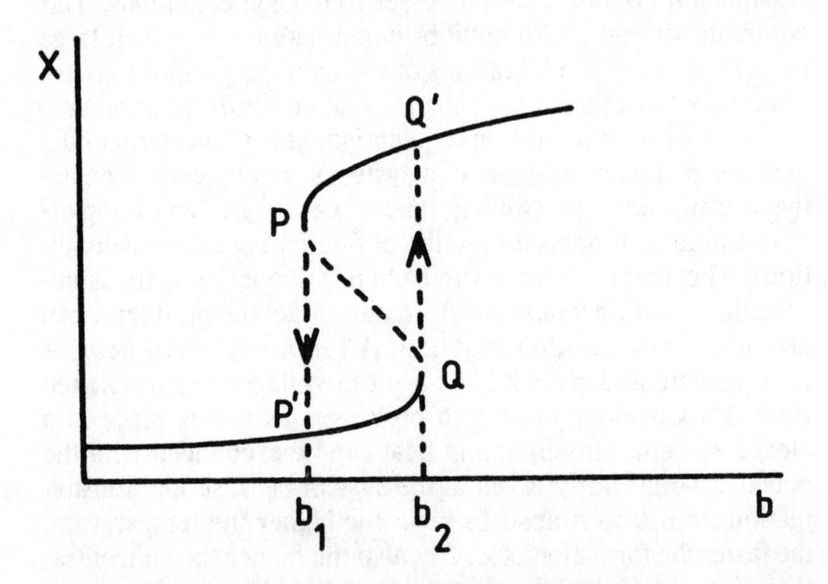

Figure 15. This figure shows how a "hysteresis" phenomenon occurs if we have the value of the bifurcation parameter b first growing and then diminishing. If the system is initially in a stationary state belonging to the lower branch, it will stay there while b grows. But at $b=b_2$, there will be a discontinuity: The system jumps from Q to Q', on the higher branch. Inversely, starting from a state on the higher branch, the system will remain there till $b=b_1$, when it will jump down to P'. Such types of bistable behavior are observed in many fields, such as lasers, chemical reactions or biological membranes.

ered as unavoidable in any "natural system." For example, in biological or ecological systems the parameters defining interaction with the environment cannot generally be considered as constants. Both the cell and the ecological niche draw their sustenance from their environment; and humidity, pH, salt concentration, light, and nutrients form a permanently fluctuating environment. The sensitivity of nonequilibrium states, not only to fluctuations produced by their internal activity but also to those coming from their environment, suggests new perspectives for biological inquiry.

Cascading Bifurcations and the Transitions to Chaos

The preceding paragraph dealt only with the first bifurcation or, as mathematicians put it, the primary bifurcation, which occurs when we push a system beyond the threshold of stability. Far from exhausting the new solutions that may appear, this primary bifurcation introduces only a single characteristic time (the period of the limit cycle) or a single characteristic length. To generate the complex spatial temporal activity observed in chemical or biological systems, we have to follow the bifurcation diagram farther.

We have already alluded to phenomena arising from the complex interplay of a multitude of frequences in hydrodynamical or chemical systems. Let us consider Bénard structures, which appear at a critical distance from equilibrium. Farther away from thermal equilibrium the convection flow begins to oscillate in time; as the distance from equilibrium is increased still farther, more and more oscillation frequencies appear, and eventually the transition to chaos is complete.[13] The interplays among the frequencies produce possibilities of large fluctuations; the "region" in the bifurcation diagram defined by such values of the parameters is often called "chaotic." In cases such as the Bénard instability, order or coherence is sandwiched between thermal chaos and nonequilibrium turbulent chaos. Indeed, if we continue to increase the gradient of temperature, the convection patterns become more complex; oscillations set in, and the ordered as-

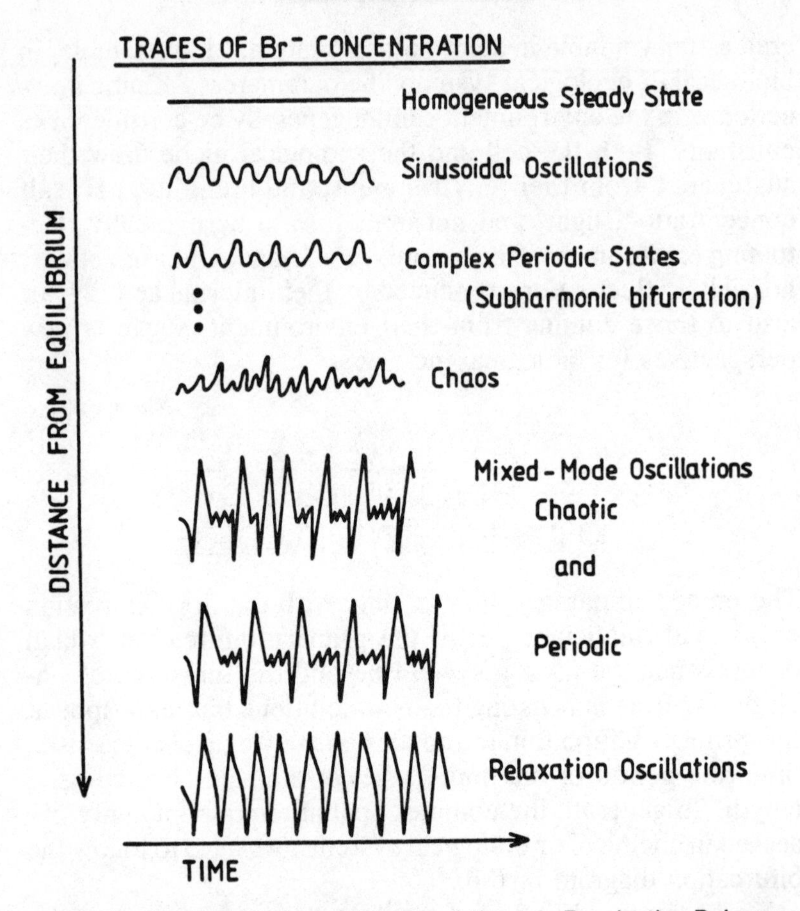

Figure 16. Temporal oscillations of the ion Br^- in the Belousov-Zhabotinski reaction. The figure represents a succession of regions corresponding to qualitative differences. This is a schematic representation. The experimental data indicate the existence of much more complicated sequences.

pect of the convection is largely destroyed. However, we should not confuse "equilibrium thermal chaos" and "nonequilibrium turbulent chaos." In thermal chaos as realized in equilibrium, all characteristic space and time scales are of molecular range, while in turbulent chaos we have such an abundance of macroscopic time and length scales that the system appears chaotic. In chemistry the relation between order and chaos appears highly complex: successive regimes of ordered (oscillatory) situations follow regimes of chaotic behavior. This has, for instance, been observed as a function of the flow rate in the Belousov-Zhabotinsky reaction.

In many cases it is difficult to disentangle the meaning of words such as "order" and "chaos." Is a tropical forest an ordered or a chaotic system? The history of any particular animal species will appear very contingent, dependent on other species and on environmental accidents. Nevertheless, the feeling persists that, as such, the overall pattern of a tropical forest, as represented, for instance, by the diversity of species, corresponds to the very archetype of order. Whatever the precise meaning we will eventually give to this terminology, it is clear that in some cases the succession of bifurcations forms an irreversible evolution where the determinism of characteristic frequencies produces an increasing randomness stemming from the multiplicity of those frequencies.

A remarkably simple road to "chaos" that has already attracted a lot of attention is the "Feigenbaum sequence." It concerns any system whose behavior is characterized by a very general feature—that is, for a determined range of parameter values the system's behavior is periodic, with a period *T;* beyond this range, the period becomes 2*T,* and beyond yet another critical threshold, the system needs 4*T* in order to repeat itself. The system is thus characterized by a succession of bifurcations, with successive period doubling. This constitutes a typical route going from simple periodic behavior to the complex aperiodic behavior occurring when the period has doubled *ad infinitum.* This route, as Feigenbaum discovered, is characterized by *universal numerical* features independent of the mechanism involved as long as the system possesses the qualitative property of period doubling. "In fact, most measurable properties of *any* such system in this aperiodic limit now can be determined in a way that essentially bypasses the details of the equations governing each specific system. . . ."[14]

In other cases, such as those represented in Figure 16, both deterministic and stochastic elements characterize the history of the system.

If we consider Figure 17 and a value of the control parameter of the order of λ_6, we see that the system already has a wealth of possible stable and unstable behaviors. The "historical" path along which the system evolves as the control parameter grows is characterized by a succession of stable regions, where deterministic laws dominate, and of instable ones, near the bifurcation points, where the system can "choose" between or among more than one possible future. Both the deterministic character of the kinetic equations whereby the set of possible states and their respective stability can be calculated, and the random fluctuations "choosing" between or among the states around bifurcation points are inextricably connected. This mixture of necessity and chance constitutes the history of the system.

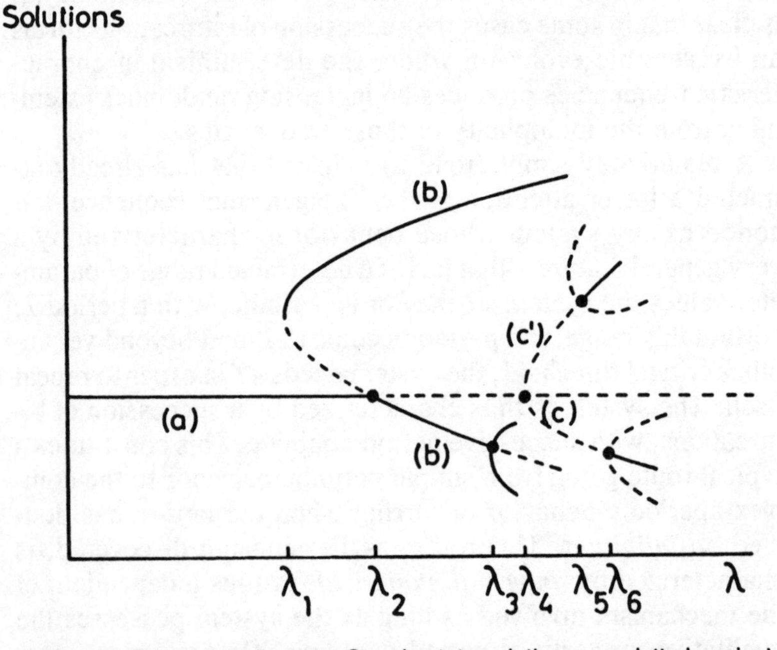

Figure 17. Bifurcation diagram. Steady-state solutions are plotted against bifurcation parameter λ. For $\lambda<\lambda_1$ there is only one stationary state for each value of λ; this set of states forms the branch a. For $\lambda=\lambda_1$ two other sets of stationary states become possible (branches b and b′).

The states of b′ are unstable but become stable at $\lambda=\lambda_2$ while the states of branch a become unstable. For $\lambda=\lambda_3$ the branch b′ is unstable again, and two other stable branches appear.

For $\lambda=\lambda_4$ the unstable branch a attains a new bifurcation point where two new branches become possible, which will be unstable up to $\lambda=\lambda_5$ and $\lambda=\lambda_6$.

From Euclid to Aristotle

One of the most interesting aspects of dissipative structures is their coherence. The system behaves as a whole, as if it were the site of long-range forces. In spite of the fact that interactions among molecules do not exceed a range of some 10^{-8} cm, the system is structured as though each molecule were "informed" about the overall state of the system.

It has often been said—and we have already repeated it—that modern science was born when Aristotelian space, for which one source of inspiration was the organization and solidarity of biological functions, was replaced by the homogeneous and isotropic space of Euclid. However, the theory of dissipative structures moves us closer to Aristotle's conception. Whether we are dealing with a chemical clock, concentration waves, or the inhomogeneous distribution of chemical products, instability has the effect of breaking symmetry, both temporal and spatial. In a limit cycle, no two instants are equivalent; the chemical reaction acquires a *phase* similar to that characterizing a light wave, for example. Again, when a favored direction results from an instability, space ceases to be isotropic. We move from Euclidian to Aristotelian space!

It is tempting to speculate that the breaking of space and time symmetry plays an important part in the fascinating phenomena of morphogenesis. These phenomena have often led to the conviction that some internal purpose must be involved, a plan realized by the embryo when its growth is complete. At the beginning of this century, German embryologist Hans Driesch believed that some immaterial "entelechy" was responsible for the embryo's development. He had discovered that the embryo at an early stage was capable of withstanding the severest perturbations and, in spite of them, of developing into a normal, functional organism. On the other hand, when we observe embryological development on film, we "see" jumps corresponding to radical reorganizations followed by periods of more "pacific" quantitative growth. There are, fortunately, few mistakes. The jumps are performed in a reproducible fashion. We might speculate that the basic mechanism of evolution is based on the play between bifurcations as mechanisms of exploration and the selection of chemical interactions stabilizing a particular trajectory. Some forty years ago, the biologist Waddington introduced such an idea. The concept of "chreod" that he introduced to describe the stabilized paths of development would correspond to possible lines of development produced as a result of the double imperative of flexibility and security.[15] Obviously the problem is very complex and can be dealt with only briefly here.

Many years ago embryologists introduced the concept of a morphogenetic field and put forward the hypothesis that the differentiation of a cell depends on its *position* in that field. But how does a cell "recognize" its position? One idea that is often debated is that of a "gradient" of a characteristic substance, of one or more "morphogens." Such gradients could actually be produced by symmetry-breaking instabilities in far-from-equilibrium conditions. Once it has been produced, a chemical gradient can provide each cell with a different chemical environment and thus induce each of them to synthesize a specific set of proteins. This model, which is now widely used, seems to be in agreement with experimental evidence. In particular, we may refer to Kauffman's work[16] on drosophila. A reaction-diffusion system is taken as responsible for the commitment to alternative development programs that appear to occur in different groups of cells in the early embryo. Each

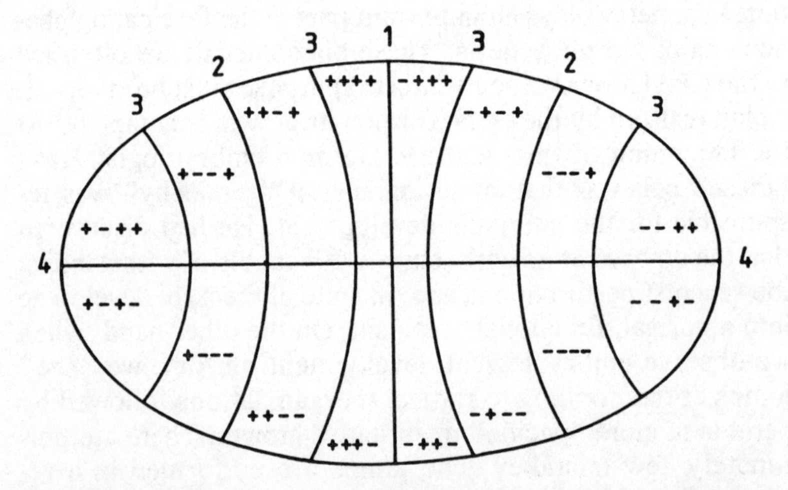

Figure 18. Schematic representation of the structure of the drosophila embryo as it results from successive binary choices. See text for more detail.

compartment would be specified by a unique combination of binary choices, each of these choices being the result of a spatial symmetry-breaking bifurcation. The model leads to successful predictions about the result of transplantations as a function of the "distance" between the original and final regions—that is, of the number of differences among the states of the binary choices or "switches" that specify each of them.

Such ideas and models are especially important in biological systems where the embryo begins to develop in an apparently symmetrical state (for example, Fucus, Acetabularia). We may ask if the embryo is really homogeneous at the beginning. And even if small inhomogeneities are present in the initial environment, do they cause or channel evolution toward a given structure? Precise answers to such questions are not available at present. However, one thing seems established: the instability connected with chemical reactions and transport appears as the only general mechanism capable of breaking the symmetry of an initially homogeneous situation.

The very possibility of such a solution takes us far beyond the age-old conflict between reductionists and antireductionists. Ever since Aristotle (and we have cited Stahl, Hegel, Bergson, and other antireductionists), the same conviction has been expressed: a concept of complex organization is required to connect the various levels of description and account for the relationship between the whole and the behavior of the parts. In answer to the reductionists, for whom the sole "cause" of organization can lie only in the part, Aristotle with his formal cause, Hegel with his emergence of Spirit in Nature, and Bergson with his simple, irrepressible, organization-creating act, assert that the whole is predominant. To cite Bergson,

> In general, when the same object appears in one aspect as simple and in another as infinitely complex, the two aspects have by no means the same importance, or rather the same degree of reality. In such cases, the simplicity belongs to the object itself, and the infinite complexity to the views we take in turning around it, to the symbols by which our senses or intellect represent it to us, or, more generally, to elements *of a different order,* with which we try to imitate it artificially, but with which it remains in-

> commensurable, being of a different nature. An artist of genius has painted a figure on his canvas. We can imitate his picture with many-coloured squares of mosaic. And we shall reproduce the curves and shades of the model so much the better as our squares are smaller, more numerous and more varied in tone. But an infinity of elements infinitely small, presenting an infinity of shades, would be necessary to obtain the exact equivalent of the figure that the artist has conceived as a simple thing, which he has wished to transport as a whole to the canvas, and which is the more complete the more it strikes us as the projection of an indivisible intuition.[17]

In biology, the conflict between reductionists and antireductionists has often appeared as a conflict between the assertion of an external and an internal purpose. The idea of an immanent organizing intelligence is thus often opposed by an organizational model borrowed from the technology of the time (mechanical, heat, cybernetic machines), which immediately elicits the retort: "Who" built the machine, the automaton that obeys external purpose?

As Bergson emphasized at the beginning of this century, both the technological model and the vitalist idea of an internal organizing power are expressions of an inability to conceive evolutive organization without immediately referring it to some preexisting goal. Today, in spite of the spectacular success of molecular biology, the conceptual situation remains about the same: Bergson's argument could be applied to contemporary metaphors such as "organizer," "regulator," and "genetic program." Unorthodox biologists such as Paul Weiss and Conrad Waddington[18] have rightly criticized the way this kind of qualification attributes to individual molecules the power to produce the global order biology aims to understand, and, by so doing, mistakes the formulation of the problem for its solution.

It must be recognized that technological analogies in biology are not without interest. However, the general validity of such analogies would imply that, as in an electronic circuit, for example, there is a basic homogeneity between the description of molecular interaction and that of global behavior: The functioning of a circuit may be deduced from the nature and position of its relays; both refer to the same scale, since the relays were designed and installed by the same engineer who built the whole machine. This cannot be the rule in biology.

It is true that when we come to a biological system such as the bacterial chemotaxis, it is hard not to speak of a molecular machine consisting of receptors, sensory and regulatory processing systems, and motor response. We know of approximately twenty or thirty receptors that can detect highly specific classes of compounds and make a bacterium swim up spatial gradients of attractants or down gradients of repellents. This "behavior" is determined by the output of the processing system—that is, the switching on or off of a tumble that generates a change in the bacterium's direction.[19]

But such cases, fascinating as they are, do not tell the whole story. In fact it is tempting to see them as limiting cases, as the end products of a specific kind of selective evolution, emphasizing stability and reproducible behavior against openness and adaptability. In such a perspective, the relevance of the technological metaphor is not a matter of principle but of opportunity.

The problem of biological order involves the transition from the molecular activity to the supermolecular order of the cell. This problem is far from being solved.

Often biological order is simply presented as an improbable physical state created and maintained by enzymes resembling Maxwell's demon, enzymes that maintain chemical differences in the system in the same way as the demon maintains temperature and pressure differences. If we accept this, biology would be in the position described by Stahl. The laws of nature allow only death. Stahl's notion of the organizing action of the soul is replaced by the genetic information contained in the nucleic acids and expressed in the formation of enzymes that permit life to be perpetuated. Enzymes postpone death and the disappearance of life.

In the context of the physics of irreversible processes, the results of biology obviously have a different meaning and different implications. We know today that both the biosphere as a whole as well as its components, living or dead, exist in far-from-equilibrium conditions. In this context life, far from being outside the natural order, appears as the supreme expression of the self-organizing processes that occur.

We are tempted to go so far as to say that once the conditions for self-organization are satisfied, life becomes as predictable as the Bénard instability or a falling stone. It is a remarkable fact that recently discovered fossil forms of life appear nearly simultaneously with the first rock formations (the oldest microfossils known today date back $3.8 \cdot 10^9$ years, while the age of the earth is supposed to be $4.6 \cdot 10^9$ years; the formation of the first rocks is also dated back to $3.8 \cdot 10^9$ years). The early appearance of life is certainly an argument in favor of the idea that life is the result of spontaneous self-organization that occurs whenever conditions for it permit. However, we must admit that we remain far from any quantitative theory.

To return to our understanding of life and evolution, we are now in a better position to avoid the risks implied by any denunciation of reductionism. A system far from equilibrium may be described as organized not because it realizes a plan alien to elementary activities, or transcending them, but, on the contrary, because the amplification of a microscopic fluctuation occurring at the "right moment" resulted in favoring one reaction path over a number of other equally possible paths. Under certain circumstances, therefore, the role played by individual behavior can be decisive. More generally, the "overall" behavior cannot in general be taken as dominating in any way the elementary processes constituting it. Self-organization processes in far-from-equilibrium conditions correspond to a delicate interplay between chance and necessity, between fluctuations and deterministic laws. We expect that near a bifurcation, fluctuations or random elements would play an important role, while between bifurcations the deterministic aspects would become dominant. These are the questions we now need to investigate in more detail.

CHAPTER VI

ORDER THROUGH FLUCTUATIONS

Fluctuations and Chemistry

In our Introduction we noted that a reconceptualization of the physical sciences is occurring today. They are moving from deterministic, reversible processes to stochastic and irreversible ones. This change of perspective affects chemistry in a striking way. As we have seen in Chapter V, chemical processes, in contrast to the trajectories of classical dynamics, correspond to irreversible processes. Chemical reactions lead to entropy production. On the other hand, classical chemistry continues to rely on a deterministic description of chemical evolution. As we have seen in Chapter V, it is necessary to produce differential equations involving the concentration of the various chemical components. Once we know these concentrations at some initial time (as well as at appropriate boundary conditions when space-dependent phenomena such as diffusion are involved), we may calculate what the concentration will be at a later time. It is interesting to note that the deterministic view of chemistry fails when far-from-equilibrium processes are involved.

We have repeatedly emphasized the role of fluctuations. Let us summarize here some of the more striking features. Whenever we reach a bifurcation point, deterministic description breaks down. The type of fluctuation present in the system will lead to the choice of the branch it will follow. Crossing a bifurcation is a stochastic process, such as the tossing of a coin. Chemical chaos provides another example (see Chapter V). Here we can no longer follow an individual chemical trajectory. We cannot predict the details of temporal evolution.

Once again, only a statistical description is possible. The existence of an instability may be viewed as the result of a fluctuation that is first localized in a small part of the system and then spreads and leads to a new macroscopic state.

This situation alters the traditional view of the relation between the microscopic level as described by molecules or atoms and the macroscopic level described in terms of global variables such as concentration. In many situations fluctuations correspond only to small corrections. As an example, let us take a gas composed of N molecules enclosed in a vessel of volume V. Let us divide this volume into two equal parts. What is the number of particles X in one of these two parts? Here the variable X is a "random" variable, and we would expect it to have a value in the neighborhood of $N/2$.

A basic theorem in probability theory, the law of large numbers, provides an estimate of the "error" due to fluctuations. In essence, it states that if we measure X we have to expect a value of the order $N/2 \pm \sqrt{N/2}$. If N is large, the difference introduced by fluctuations $\sqrt{N/2}$ may also be large (if $N = 10^{24}$, $\sqrt{N} = 10^{12}$); however, the relative error introduced by fluctuations is of the order of $(\sqrt{N/2})/(N/2)$ or $1/\sqrt{N}$ and thus tends toward zero for a sufficiently large value of N. As soon as the system becomes large enough, the law of large numbers enables us to make a clear distinction between mean values and fluctuations, and the latter may be neglected.

However, in nonequilibrium processes we may find just the opposite situation. Fluctuations determine the global outcome. We could say that instead of being corrections in the average values, fluctuations now modify those averages. This is a new situation. For this reason we would like to introduce a neologism and call situations resulting from fluctuation "order through fluctuation." Before giving examples, let us make some general remarks to illustrate the conceptual novelty of this situation.

Readers may be familiar with the Heisenberg uncertainty relations, which express in a striking way the probabilistic aspects of quantum theory. Since we can no longer simultaneously measure position and momentum in quantum theory, classical determinism is breaking down. This was believed to be of no importance for the description of macroscopic objects such as living systems. But the role of fluctuations in nonequilibrium systems shows that this is not the case. Randomness remains essential on the macroscopic level as well. It is interesting to note another analogy with quantum theory, which assigns a wave behavior to all elementary particles. As we have seen, chemical systems far from equilibrium may also lead to coherent wave behavior: these are the chemical clocks discussed in Chapter V. Once again, some of the properties quantum mechanics discovered on the microscopic level now appear on the macroscopic level.

Chemistry is actively involved in the reconceptualization of science.[1] We are probably only at the beginning of new directions of research. It may well be, as some recent calculations suggest, that the idea of reaction rate has to be replaced in some cases by a statistical theory involving a distribution of reaction probabilities.[2]

Fluctuations and Correlations

Let us go back to the types of chemical reaction discussed in Chapter V. To take a specific example, consider a chain of reactions such as $A \rightleftarrows X \rightleftarrows F$. The kinetic equations in Chapter V refer to the average concentrations. To emphasize this we shall now write $\langle X \rangle$ instead of X. We can then ask what is the probability at a given time of finding a number X for the concentration of this component. Obviously this probability will fluctuate, as do the number of collisions among the various molecules involved. It is easy to write an equation that describes the change in this probability distribution $P(X,t)$ as a result of processes that produce molecule X and of processes that destroy that molecule. We may perform the calculation for equilibrium systems or for steady-state systems. Let us first mention the results obtained for equilibrium systems.

At equilibrium we virtually recover a classical probabilistic distribution, the Poisson distribution, which is described in every textbook on probabilities, since it is valid in a variety of situations, such as the distribution of telephone calls, waiting times in restaurants, or the fluctuation of the concentration of particles in a gas or a liquid. The mathematical form of this distribution is of no importance here. We merely want to emphasize two of its aspects. First, it leads to the law of large numbers as formulated in the first section of this chapter. Thus fluctuations indeed become negligible in a large system. Moreover, this law enables us to calculate the correlation between the number of particles X at two different points in space separated by some distance r. The calculation demonstrates that at equilibrium there is no such correlation. The probability of finding two molecules X and X' at two different points r and r' is the product of finding X at r and X' at r' (we consider distances that are large in respect to the range of intermolecular forces).

One of the most unexpected results of recent research is that this situation changes drastically when we move to nonequilibrium situations. First, when we come close to bifurcation points the fluctuations become abnormally high and the law of large numbers is violated. This is to be expected, since the system may then "choose" among various regimes. Fluctuations can even reach the same order of magnitude as the mean macroscopic values. Then the distinction between fluctuations and mean values breaks down. Moreover, in the case of a nonlinear type of chemical reaction discussed in Chapter V, long-range correlations appear. Particles separated by macroscopic distances become linked. Local events have repercussions throughout the whole system. It is interesting to note[3] that such long-range correlations appear at the precise point of transition from equilibrium to nonequilibrium. From this point of view the transition resembles a phase transition. However, the amplitudes of these long-range correlations are at first small but increase with distance from equilibrium and may become infinite at the bifurcation points.

We believe that this type of behavior is quite interesting, since it gives a molecular basis to the problem of communication mentioned in our discussion of the chemical clock. Even before the macroscopic bifurcation, the system is organized through these long-range correlations. We come back to one of the main ideas of this book: nonequilibrium as a source of order. Here the situation is especially clear. At equilibrium molecules behave as essentially independent entities; they ignore one another. We would like to call them "hypnons," "sleepwalkers." Though each of them may be as complex as we like, they ignore one another. However, nonequilibrium wakes them up and introduces a coherence quite foreign to equilibrium. The microscopic theory of irreversible processes that we shall develop in Chapter IX will present a similar picture of matter.

Matter's activity is related to the nonequilibrium conditions that it itself may generate. Just as in macroscopic behavior, the laws of fluctuations and correlations are universal at equilibrium (when we find the Poisson type of distribution); they become highly specific depending on the type of nonlinearity involved when we cross the boundary between equilibrium and nonequilibrium.

The Amplification of Fluctuations

Let us first take two examples wherein the growth of a fluctuation preceding the formation of a new structure can be followed in detail. The first is the aggregation of slime molds, which when threatened with starvation coalesce into a single supracellular mass. We have already mentioned this in Chapter V. Another illustration of the role of fluctuations is the first stage in the construction of a termites' nest. This was first described by Grassé, and Deneubourg has studied it from the standpoint that interests us here.[4]

Self-Aggregation Process in an Insect Population

Larvae of a coleoptera (*Dendroctonus micans* [*Scol.*]). are initially distributed at random between two horizontal sheets of glass, 2 mm apart. The borders are open and the surface is equal to 400 cm^2.

The aggregation process appears to result from the competition between two factors: the random moves of the larvae, and their reaction to a chemical product, a "pheromon" they synthetize from terpenes contained in the tree on which they feed and that each of them emits at a rate depending on its nutrition state. The pheromon diffuses in space, and the larvae move in the direction of its concentration gradient. Such a reaction provides an autocatalytic mechanism since, as they gather in a cluster, the larvae contribute to enhance the attractiveness of the corresponding region. The higher the local density of larvae in this region, the stronger the gradient and the more intense the tendency to move toward the crowded point.

The experiment shows that the density of the larvae population determines not only the rate of the aggregation process but its effectiveness as well—that is, the number of larvae that will finally be part of the cluster. At high density (Figure A) a cluster appears and rapidly grows at the center of the experimental setup. At very low density (Figure B), no stable cluster appears.

Moreover, other experiments have explored the possibility for a cluster to develop starting from a "nucleus" artificially created in a peripheral region of the system. Different solutions appear depending on the number of larvae in this initial nucleus.

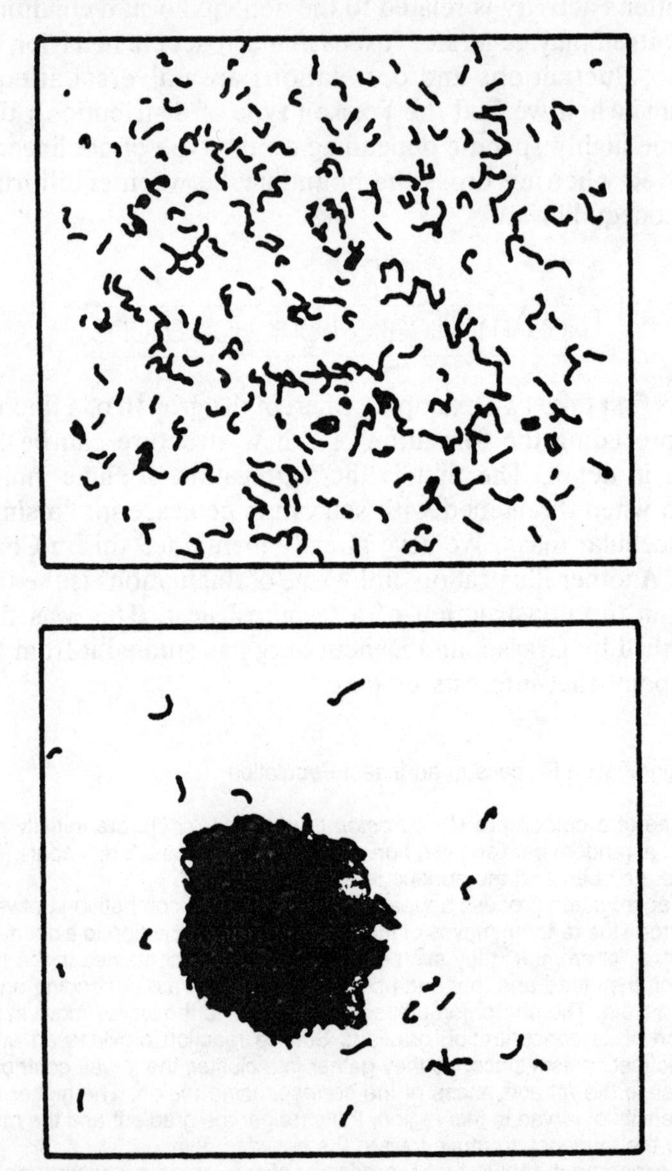

Figure A. Self-aggregation at high density. The times are 0 minutes and 21 minutes.

If this number is small compared with the total number of larvae, the cluster fails to develop (Figure D). If it is large, the cluster grows (Figure E). For intermediate values of the initial nucleus, new types of structure may develop: Two, three or four other clusters appear and coexist, with a time of life at least greater than the time of observation (Figures F and G).

No such multicluster structure was ever observed in experiments with homogeneous initial conditions. It would seem they correspond, in a bifurcation

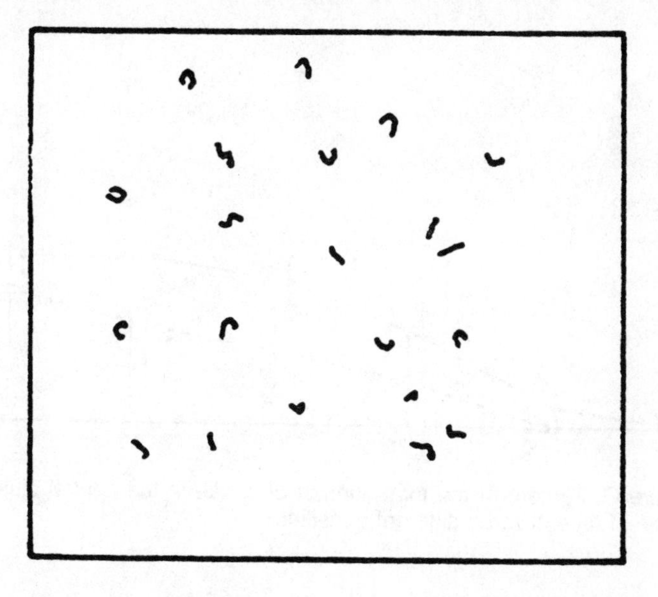

Figure B. Self-aggregation at low density. The times are 0 minutes and 22 minutes.

diagram, to stable states compatible with the value of the parameters characterizing the system but that cannot be attained by this system starting from homogeneous conditions. The nucleus would play the part of a finite perturbation necessary to excite the system and deport it in a region of the bifurcation diagram corresponding to such families of multicluster solutions.

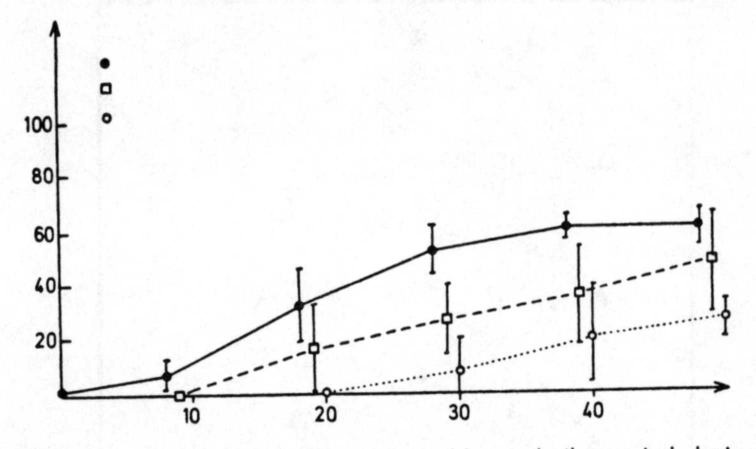

Figure C. Percent of the total number of larvae in the central cluster in function of time at three different densities.

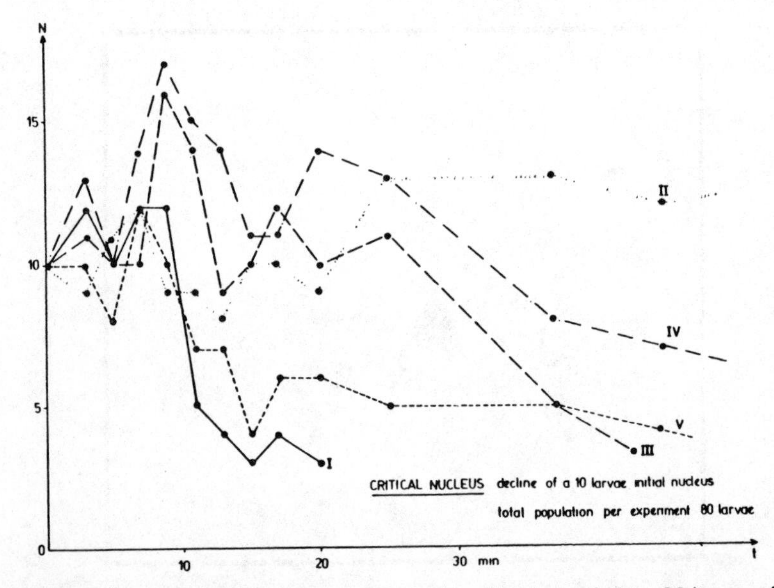

Figure D. Fall of initial clusters of 10 larvae. Total population, 80 larvae. *N:* number of larvae in clusters.

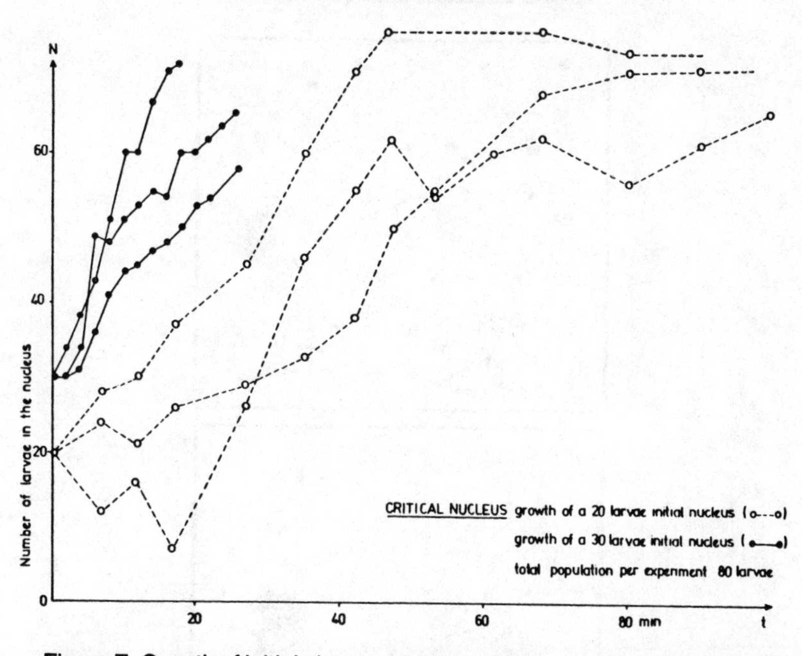

Figure E. Growth of initial clusters of 20 and 30 larvae. Total population, 80 larvae.

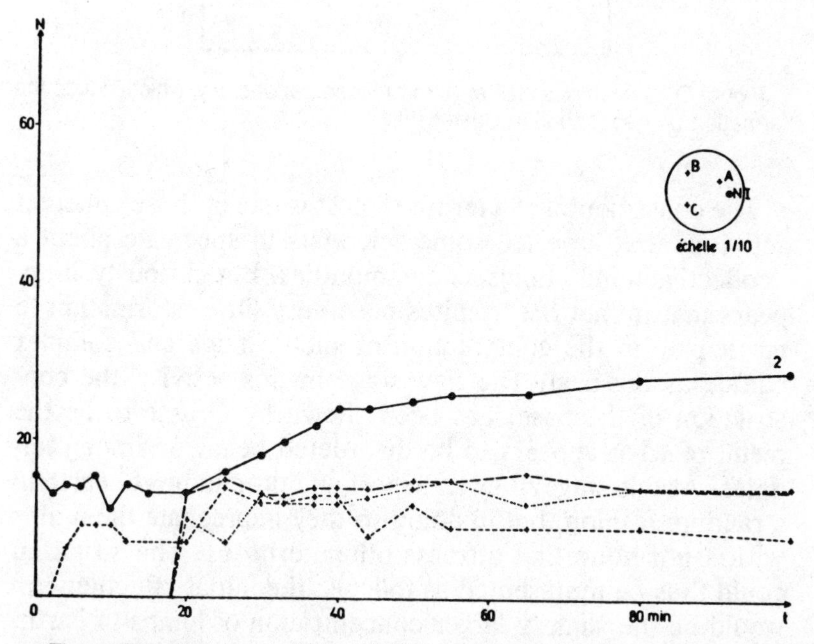

Figure F. Multicluster solutions. Initial value of the cluster, 15 larvae. Total population, 80 larvae.

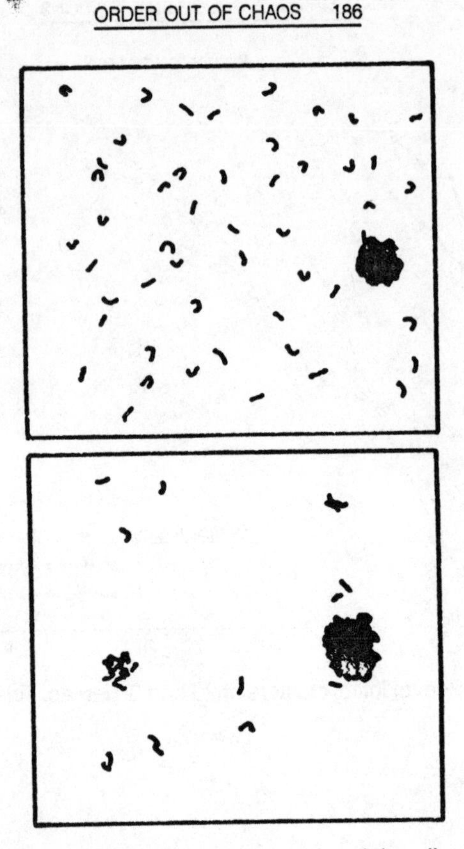

Figure G. Growth of a cluster (I) introduced peripherally, which induce the formation of a second little cluster (II).

The construction of a termites' nest is one of those coherent activities that have led some scientists to speculate about a "collective mind" in insect communities. But curiously, it appears that in fact the termites need very little information to participate in the construction of such a huge and complex edifice as the nest. The first stage in this activity, the construction of the base, has been shown by Grassé to be the result of what appears to be disordered behavior among termites. At this stage, they transport and drop lumps of earth in a random fashion, but in doing so they impregnate the lumps with a hormone that attracts other termites. The situation could thus be represented as follows: the initial "fluctuation" would be the slightly larger concentration of lumps of earth, which inevitably occurs at one time or another at some point in the area. The amplification of this event is produced by the increased density of termites in the region, attracted by the slightly higher hormone concentration. As termites become more numerous in a region, the probability of their dropping lumps of earth there increases, leading in turn to a still higher concentration of the hormone. In this way "pillars" are formed, separated by a distance related to the range over which the hormone spreads. Similar examples have recently been described.

Although Boltzmann's order principle enables us to describe chemical or biological processes in which differences are leveled out and initial conditions forgotten, it cannot explain situations such as these, where a few "decisions" in an unstable situation may channel a system formed by a large number of interactive entities toward a global structure.

When a new structure results from a finite perturbation, the fluctuation that leads from one regime to the other cannot possibly overrun the initial state in a single move. It must first establish itself in a limited region and then invade the whole space: there is a *nucleation* mechanism. Depending on whether the size of the initial fluctuating region lies below or above some critical value (in the case of chemical dissipative structures, this threshold depends in particular on the kinetic constants and diffusion coefficients), the fluctuation either regresses or else spreads to the whole system. We are familiar with nucleation phenomena in the classical theory of phase change: in a gas, for example, condensation droplets incessantly form and evaporate. That temperature and pressure reach a point where the liquid state becomes stable means that a critical droplet size can be defined (which is smaller the lower the temperature and the higher the pressure). If the size of a droplet exceeds this "nucleation threshold," the gas almost instantaneously transforms into a liquid.

Moreover, theoretical studies and numerical simulations show that the critical nucleus size increases with the efficacy of the diffusion mechanisms that link all the regions of systems. In other words, the faster communication takes place within a system, the greater the percentage of unsuccessful fluctuations and thus the more stable the system. This aspect of the critical-size problem means that in such situations the "outside world," the environment of the fluctuating region, always tends to damp fluctuations. These will be destroyed or

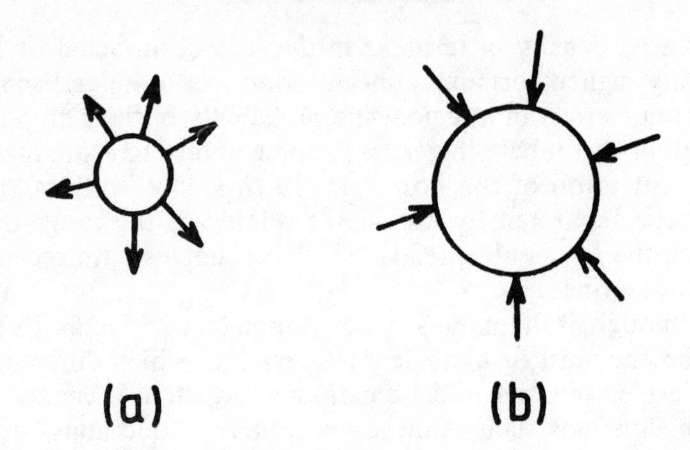

Figure 19. Nucleation of a liquid droplet in a supersaturated vapor. (a) droplet smaller than the critical size; (b) droplet larger than the critical size. The existence of the threshold has been experimentally verified for dissipative structures.

amplified according to the effectiveness of the communication between the fluctuating region and the outside world. The critical size is thus determined by the competition between the system's "integrative power" and the chemical mechanisms amplifying the fluctuation.

This model applies to the results obtained recently in *in vitro* experimental studies of the onset of cancer tumors.[5] An individual tumor cell is seen as a "fluctuation," uncontrollably and permanently able to appear and to develop through replication. It is then confronted with the population of cytotoxic cells that either succeeds in destroying it or fails. Following the values of the different parameters characteristic of the replication and destruction processes, we can predict a regression or an amplification of the tumor. This kind of kinetic study has led to the recognition of unexpected features in the interaction between cytotoxic cells and the tumor. It seems that cytotoxic cells can confuse dead tumor cells with living ones. As a result, the destruction of the cancer cells becomes increasingly difficult.

The question of the limits of complexity has often been raised. Indeed, the more complex a system is, the more numerous are the types of fluctuations that threaten its stability. How then, it has been asked, can systems as complex as ecological or human organizations possibly exist? How do they manage to avoid permanent chaos? The stabilizing effect of communication, of diffusion processes, could be a partial answer to these questions. In complex systems, where species and individuals interact in many different ways, diffusion and communication among various parts of the system are likely to be efficient. There is competition between stabilization through communication and instability through fluctuations. The outcome of that competition determines the threshold of stability.

Structural Stability

When can we begin to speak about "evolution" in its proper sense? As we have seen, dissipative structures require far-from-equilibrium conditions. Yet the reaction diffusion equations contain parameters that can be shifted back to near-equilibrium conditions. The system can explore the bifurcation diagram in both directions. Similarly, a liquid can shift from laminar flow to turbulence and back. There is no definite evolutionary pattern involved.

The situation for models involving the size of the system as a bifurcation parameter is quite different. Here, growth occurring irreversibly in time produces an irreversible evolution. But this remains a special case, even if it can be relevant for morphogenetic development.

Be it in biological, ecological, or social evolution, we cannot take as given either a definite set of interacting units, or a definite set of transformations of these units. The definition of the system is thus liable to be modified by its evolution. The simplest example of this kind of evolution is associated with the concept of structural stability. It concerns the reaction of a given system to the introduction of new units able to multiply by taking part in the system's processes.

The problem of the stability of a system vis-à-vis this kind of change may be formulated as follows: the new constituents, introduced in small quantities, lead to a new set of reactions among the system's components. This new set of reactions then enters into competition with the system's previous mode of functioning. If the system is "structurally stable" as far as this intrusion is concerned, the new mode of functioning will be unable to establish itself and the "innovators" will not survive. If, however, the structural fluctuation successfully imposes itself—if, for example, the kinetics whereby the "innovators" multiply is fast enough for the latter to invade the system instead of being destroyed—the whole system will adopt a new mode of functioning: its activity will be governed by a new "syntax."[6]

The simplest example of this situation is a population of macromolecules reproduced by polymerization inside a system being fed with the monomers *A* and *B*. Let us assume the polymerization process to be autocatalytic—that is, an already synthesized polymer is used as a model to form a chain having the same sequence. This kind of synthesis is much faster than a synthesis in which there is no model to copy. Each type of polymer, characterized by a particular sequence of *A* and *B*, can be described by a set of parameters measuring the speed of the synthesis of the copy it catalyzes, the accuracy of the copying process, and the mean life of the macromolecule itself. It may be shown that, under certain conditions, a single type of polymer having a sequence, shall we say, *ABABABA* . . . dominates the population, the other polymers being reduced to mere "fluctuations" with respect to the first. The problem of structural stability arises each time that, as a result of a copying "error," a new type of polymer characterized by a hitherto unknown sequence and by a new set of parameters appears in the system and begins to multiply, competing with the dominant species for the available *A* and *B* monomers. Here we encounter an elementary case of the classic Darwinian idea of the "survival of the fittest."

Such ideas form the basis for the model of prebiotic evolution developed by Eigen and his coworkers. The details of Eigen's argument are easily accessible elsewhere.[7] Let us briefly state that it seems to show that there is only one type of system that can resist the "errors" that autocatalytic populations continually make—a polymer system structurally stable for any possible "mutant polymer." This system is composed of two sets of polymer molecules. The molecules of the first set are of the "nucleic acid" type; each molecule is capable of reproducing itself and acts as a catalyst in the synthesis of a molecule of the second set, which is of the proteic type: each molecule of this second set catalyzes the self-reproduction of a molecule of the first set. This transcatalytic association between molecules of the two sets may turn into a cycle (each "nucleic acid" reproduces itself with the help of a "protein"). It is then capable of stable survival, sheltered from the continual emergence of new polymers with higher reproductive efficiency: indeed, nothing can intrude into the self-replicating cycle formed by "proteins" and "nucleic acids." A new kind of evolution may thus begin to grow on this stable foundation, heralding the genetic code.

Eigen's approach is certainly of great interest. Darwinian selection for faithful self-reproduction is certainly important in an environment with a limited capacity. But we tend to believe that this is not the only aspect involved in prebiotic evolution. The "far-from-equilibrium" conditions related to critical amounts of flow of energy and matter are also important. It seems reasonable to assume that some of the first stages moving toward life were associated with the formation of mechanisms capable of absorbing and transforming chemical energy, so as to push the system into "far-from-equilibrium" conditions. At this stage life, or "prelife," probably was so diluted that Darwinian selection did not play the essential role it did in later stages.

Much of this book has centered around the relation between the microscopic and the macroscopic. One of the most important problems in evolutionary theory is the eventual feedback between macroscopic structures and microscopic events: macroscopic structures emerging from microscopic events would in turn lead to a modification of the microscopic mechanisms. Curiously, at present, the better understood cases concern social situations. When we build a road or a bridge, we can predict how this will affect the behavior of the population, and this will in turn determine other modifications of the modes of communication in the region. Such interrelated processes generate very complex situations, the understanding of which is needed before any kind of modelization. This is why what we will now describe are only very simple cases.

Logistic Evolution

In social cases, the problem of structural stability has a large number of applications. But it must be emphasized that such applications imply a drastic simplification of a situation defined simply in terms of competition between self-replicating processes in an environment where only a limited amount of the needed resources exists.

In ecology the classic equation for such a problem is called the "logistic equation." This equation describes the evolution of a population containing N individuals, taking into account the birthrate, the deathrate, and the amount of resources available to the population. The logistic equation can be written $dN/dt = r'N(1-N/K)-mN$, where r and m are characteristic birth and death constants and K the "carrying capacity" of the environment. Whatever the initial value of N, as time goes on it will reach the steady-state value $N = K - m/r$ determined by the differences of the carrying capacity and the ratio of death and birth constants. When this value is reached, the environ-

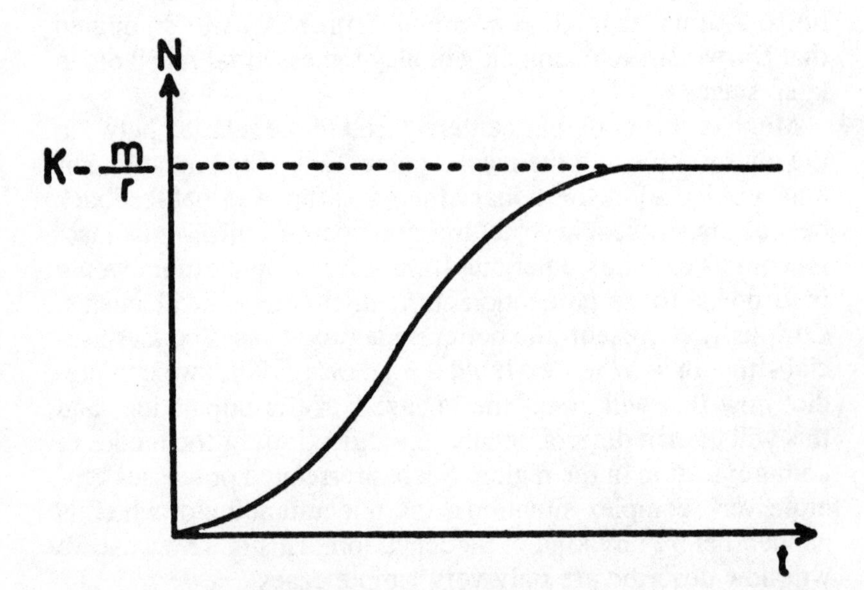

Figure 20. Evolution of a population N as a function of time t according to the logistic curve. The stationary state $N=0$ is unstable while the stationary state $N=K-m/r$ is stable with respect to fluctuations of N.

ment is saturated, and at each instant as many individuals die as are born.

The apparent simplicity of the logistic equation conceals to some extent the complexity of the mechanisms involved. We have already mentioned the effect of external noise, for example. Here it has an especially simple meaning. Obviously, if only because of climatic fluctuations, the coefficients K, m, and r cannot be taken as constant. We know that such fluctuations can completely upset the ecological equilibrium and even drive the population to extinction. Of course, as a result, new processes, such as the storage of food and the formation of new colonies, will begin and eventually evolve so that some effects of external fluctuation may be avoided.

But there is more. Instead of writing the logistic equation as continuous in time, let us compare the population at fixed time intervals (for example, separated by a year). This "discrete" logistic equation can be written in the form $N_{t+1} = N_t(1 + rK[1 - N_t/K])$, where N_t and N_{t+1} are the populations separated by a one-year interval (we neglect here the death term). The remarkable feature, noted by R. May,[8] is that such equations, in spite of their simplicity, admit a bewildering number of solutions. For values of the parameter $0 \leq r \leq 2$, we have, as in the continuous case, a uniform approach to equilibrium. For values of r lower than 2.444, a limit cycle sets in: we now have a periodic behavior with a two-year period. This is followed by four-, eight-, etc., year cycles, until the behavior can only be described as chaotic (if r is larger than 2.57). Here we have a transition to chaos as described in Chapter V. Does this chaos arise in nature? Recent studies[9] seem to indicate that the parameters characterizing natural populations keep them from the chaotic region. Why is this so? Here we have one of the very interesting problems created by the confluence of evolutionary problems with the mathematics produced by computer simulation.

Up to now we have taken a static point of view. Let us now move to mechanisms, whereby the parameters K, r, and m may vary during biological or ecological evolution.

We have to expect that during evolution the values of the ecological parameters K, r, and m will vary (as well as many other parameters and variables, whether they are quantifiable or not). Living societies continually introduce new ways of exploiting existing resources or of discovering new ones (that is, K increases) and continually discover new ways of extending their lives or of multiplying more quickly. Each ecological equilibrium defined by the logistic equation is thus only temporary, and a logistically defined niche will be occupied successively by a series of species, each capable of ousting the preceding one when its "aptitude" for exploiting the niche, as measured by the quantity $K-m/r$, becomes greater. (See Figure 21.) Thus the logistic equation leads to the definition of a very simple situation where we can give a quantitative formulation of the Darwinian idea of the "survival of the fittest." The "fittest" is the species for which at a given time the quantity $K-m/r$ is the largest.

As restricted as the problem described by the logistic equation is, it nonetheless leads to some marvelous examples of nature's inventiveness.

Take the example of caterpillars, who must remain undetected, since the slowness of their movement makes escape impossible.

The evolved strategies of using poisons and irritating hairs and spines, as well as intimidating displays, are highly effective in repelling birds and other potential predators. But none

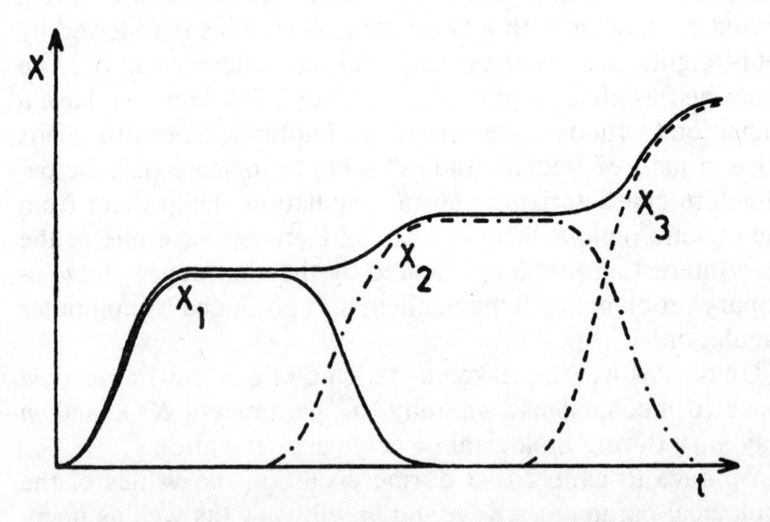

Figure 21. Evolution of total population X as function of time; the population is made up by species X_1, X_2 and X_3, which appear successively and are characterized by increasing values of $K-m/r$ (see text).

of these strategies is effective against all predators at all times, particularly if a predator is hungry enough. The ideal strategy is to remain totally undetected. Some caterpillars approach this ideal, and the variety and sophistication of the strategies used by the hundreds of lepidopteran species to remain undetected bring to mind the words of distinguished nineteenth-century naturalist Louis Agassiz: "The possibilities of existence run so deeply into the extravagant that there is scarcely any conception too extraordinary for Nature to realize."[10]

We cannot resist giving an example reported by Milton Love.[11] The sheep liver trematode has to pass from an ant to a sheep, where it will finally reproduce itself. The chances of sheep swallowing an infected ant are very small, but the ant behaves in a remarkable way: it starts to maximize the probability of its encounter with a sheep. The trematode has truly "body snatched" its host. It has burrowed into the ant's brain, compelling its victim to behave in a suicidal way: the possessed ant, instead of staying on the ground, climbs to the tip of a blade of grass and there, immobile, waits for a sheep. This is indeed an incredibly "clever" solution to the parasites problem. How it was selected remains a puzzle.

Other situations in biological evolution may be investigated using models similar to the logistic equation. For instance, it is possible to calculate the conditions of interspecies competition under which it may be advantageous for a fraction of the population to specialize in warlike and nonproductive activity (for example, the "soldiers" among the social insects). We can also determine the kind of environment in which a species that has become specialized, that has restricted the range of its food resources, will survive more easily than a nonspecialized species that consumes a wider range of resources.[12] But here we are approaching some very different problems, which concern the organization of internally differentiated populations. Clear distinctions are absolutely necessary if we are to avoid confusion. In populations where individuals are not interchangeable and where each, with its own memory, character, and experience, is called upon to play a singular role, the relevance of the logistic equation and, more generally, of any simple Darwinian reasoning becomes quite relative. We shall return to this problem.

It is interesting to note that the type of curve represented in

Figure 21 showing the succession of growths and peaks defined by a given logistic equation's family with increasing $K - m/r$ has also been used to describe the multiplication of certain technical procedures or products. Here too, the discovery or introduction of a new technique or product breaks some kind of social, technological, or economic equilibrium. This equilibrium would correspond to the maximum reached by the growth curve of the techniques or products with which the innovation is going to have to compete and that play a similar role in the situation described by the equation.[13] Thus, to choose but one example, not only did the spread of the steamship lead to the disappearance of most sailing ships, but, by reducing the cost of transportation and increasing its speed, it caused an increase in the demand for sea transport *("K")* and consequently an increase in the population of ships. We are obviously representing here an extremely simple situation, supposedly governed by purely economic logic. Indeed, in this case innovation seems merely to satisfy, albeit in a different way, a preexisting need that remains unchanged. However, in ecology as in human societies, many innovations are successful without such a preexisting "niche." Such innovations transform the environment in which they appear, and as they spread, they create the conditions necessary for their own multiplication, their "niche." In social situations, in particular, the creation of a "demand," and even of a "need" for this demand to fulfill, often appears as correlated with the production of the goods or techniques that satisfy the demand.

Evolutionary Feedback

A first step toward accounting for this dimension of the evolutionary process can be achieved by making the "carrying capacity" of a system a function of the way it is exploited instead of taking it as given.

In this way some supplementary dimensions of economic activities, and more particularly the "multiplying effects," can be represented. Thus we can describe the self-accelerating properties of systems and the spatial differentiation between different levels of activity.

Geographers have already constructed a model correlating these processes, the Christaller model, defining the optimal spatial distribution of centers of economic activity. Important centers would be at the intersection of an hexagonal network, each being surrounded by a ring of towns of the next smallest size, each being, etc. . . . Obviously, in actual cases, such a regular hierarchical distribution is very infrequent: historical, political, and geographical factors abound, disrupting the spatial symmetry. But there is more. Even if all the important sources of asymmetrical development were excluded and we started from a homogeneous economic and geographical space, the modeling of the genesis of a distribution such as defined by Christaller establishes that the kind of static optimalization he describes constitutes a possible but quite unlikely result of the process.

The model in question[14] stages only the minimal set of variables implied by a calculation such as Christaller. A set of equations extending the logistic equations is constructed, starting from the basic supposition that populations tend to migrate as a function of local levels of economic activity, which thus define a kind of local "carrying capacity," here reduced to an "employment" capacity. But the local population is also a potential consumer for locally produced goods. We have, in fact, a double positive feedback, called the "urban multiplier," for a local development: both the local population and the economic infrastructure produced by the already attained level of activity accelerate the increase of this activity. But each local level of activity is also determined by competition with similar centers of activity located elsewhere. The sale of produced goods or services depends on the cost of transporting them to consumers and on the size of the "enterprise." The expansion of each such enterprise depends on a demand that this expansion itself helps to create and for which it competes. Thus the respective growth of population and manufacturing or service activities is linked by strong feedback and nonlinearities.

The model starts with a hypothetical initial condition, where "level 1" activity (rural) exists at the different points; it then permits us to follow successive launchings of activities corresponding to "superior" levels in Christaller's hierarchy—that is, implying exportation on a greater range. Even if the initial

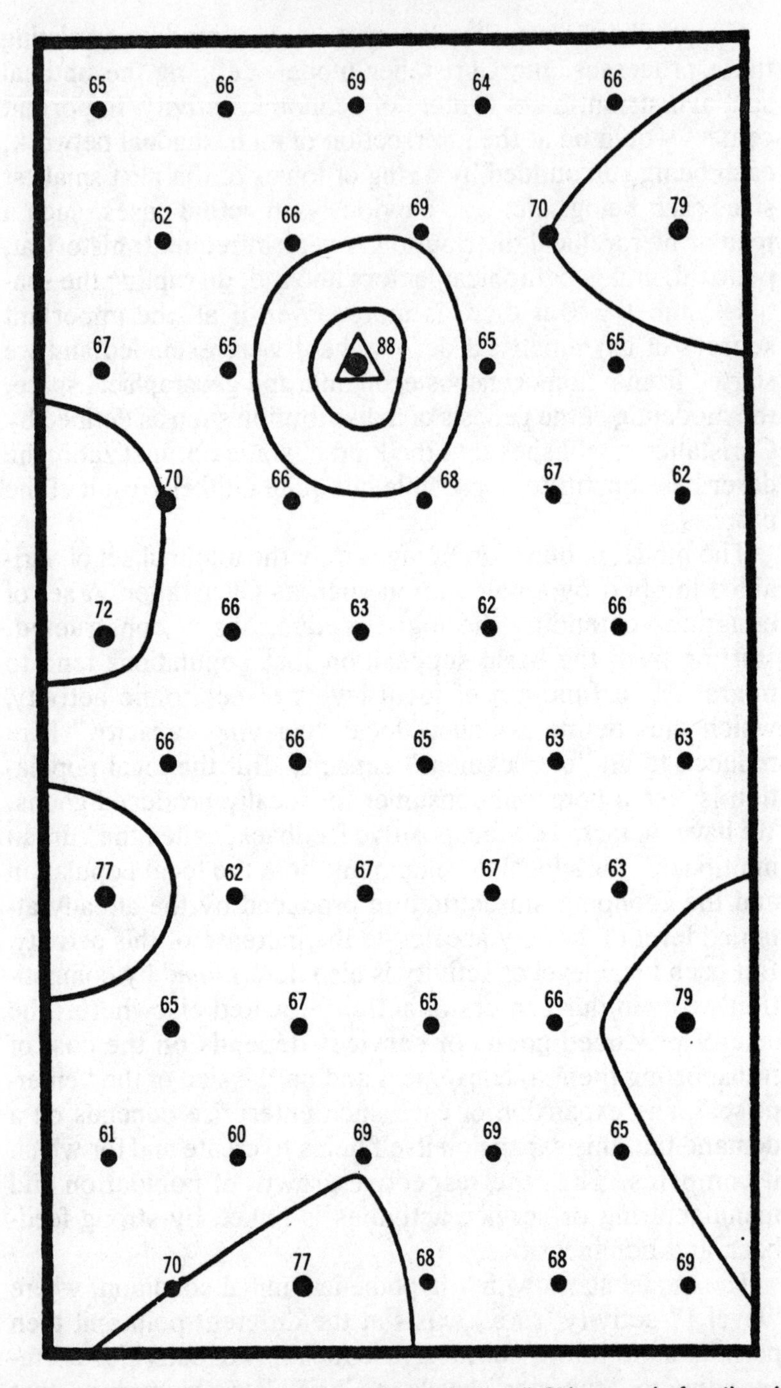

Figure 22. A possible history of "urbanization." ● have only function 1; ● have functions 1 and 2; △ have functions 1, 2 and 3. [△] are the largest centers, with functions 1, 2, 3, and 4. At $t=0$ (not represented), all points

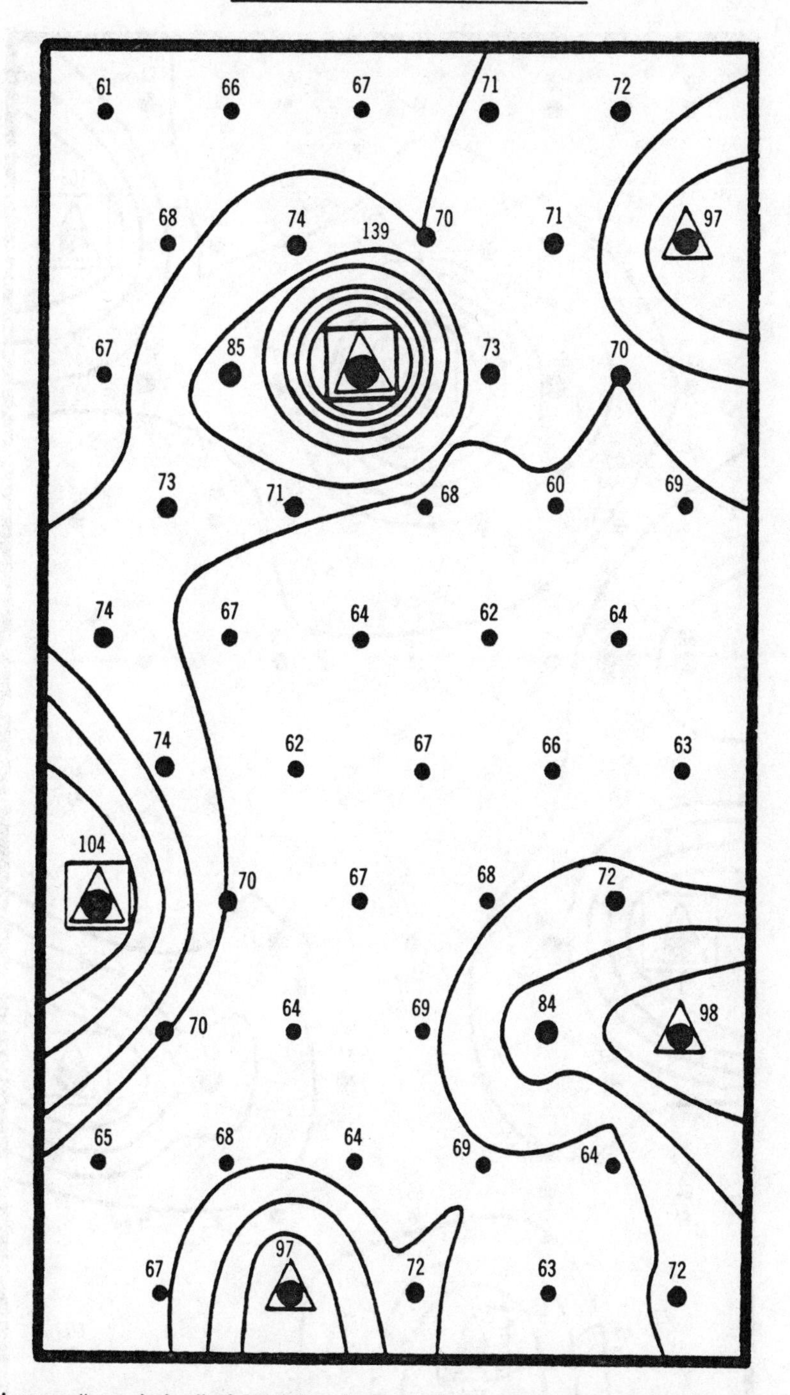

have a "population" of 67 units. At C, the largest center is going through a maximum (152 population units); this is followed by an "urban sprawl," with creation of satellite cities; this also occurs around the second main center.

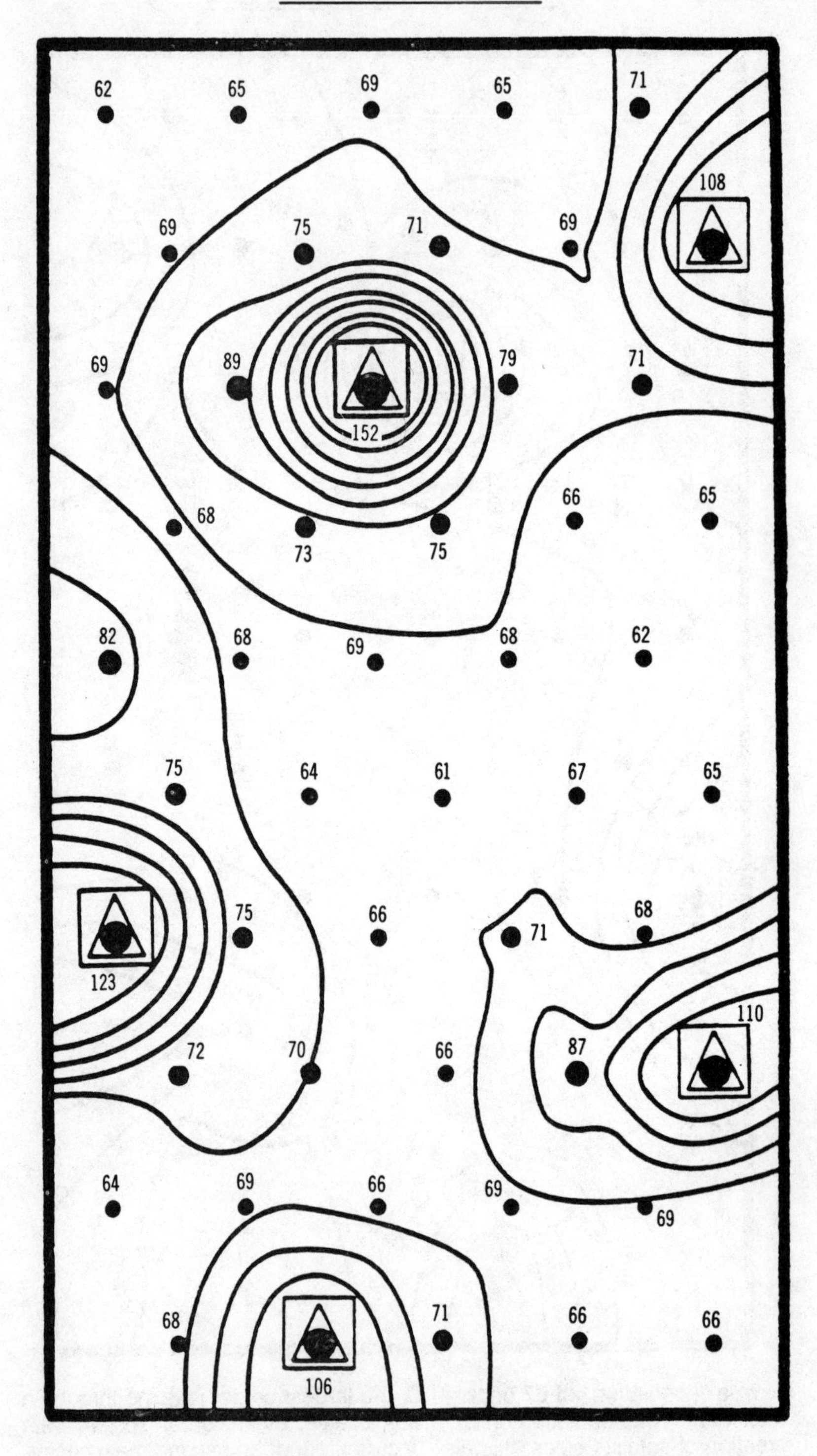
62
65
69
65
71
108
69
75
71
69
69
89
152
79
71
68
73
75
66
65
82
68
69
68
62
75
64
61
67
65
75
66
71
68
123
110
72
70
66
87
64
69
66
69
69
68
106
71
66
66

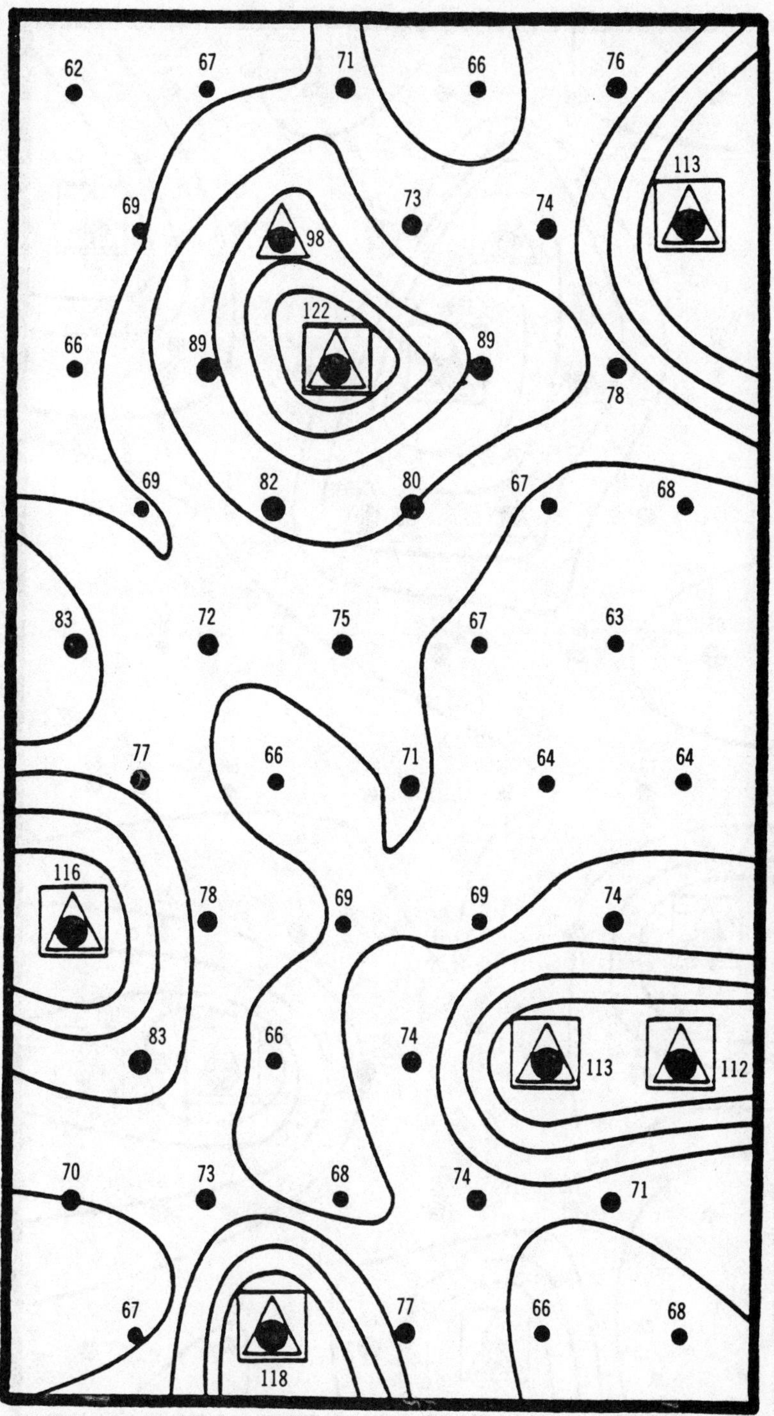

62
67
71
66
76
113
69
98
73
74
122
66
89
89
78
69
82
80
67
68
83
72
75
67
63
77
66
71
64
64
116
78
69
69
74
83
66
74
113
112
70
73
68
74
71
67
118
77
66
68

67
70
72
67
72
109
75
77
77
110
135
92
71
75
84
70
92
93
71
65
87
75
72
69
66
73
68
67
68
66
69
77
82
80
118
108
73
80
85
125
62
77
71
79
87
114
68
81
69
73

state is quite homogeneous, the model shows that the mere play of chance factors—factors uncontrolled by the model, such as the place and time where the different enterprises start—is sufficient to produce symmetry breakings: the appearance of highly concentrated zones of activity while others suffer a reduction in economic activity and are depopulated. The different computer simulations show growth and decay, capture and domination, periods of opportunity for alternative developments followed by solidification of the existing domination structures.

Whereas Christaller's symmetrical distribution ignores history, this scenario takes it into account, at least in a very minimal sense, as an interplay between "laws," in this case of a purely economic nature, and the "chance" governing the sequence of launchings.

Modelizations of Complexity

In spite of its simplicity, our model succeeds in showing some properties of the evolution of complex systems, and in particular, the difficulty of "governing" a development determined by multiple interacting elements. Each individual action or each local intervention has a collective aspect that can result in quite unanticipated global changes. As Waddington emphasized, at present we have very little understanding of how a complex system is likely to respond to a given change. Often this response runs counter to our intuition. The term "counterintuitive" was introduced at MIT to express our frustration: "The damn thing just does not do what it should do!" To take the classic example cited by Waddington, a program of slum clearance results in a situation worse than before. New buildings attract a larger number of people into the area, but if there are not enough jobs for them, they remain poor, and their dwellings become even more overcrowded.[15] We are trained to think in terms of linear causality, but we need new "tools of thought": one of the greatest benefits of models is precisely to help us discover these tools and learn how to use them.

As we have already emphasized, logistic equations are most relevant when the crucial dimension is the growth of a population, be it of animals, activities, or habits. What is presupposed is that each member of a given population can be taken as the equivalent of any of the others. But this general equivalence can itself be seen not as a simple general fact but as an approximation, the validity of which depends on the constraints and pressures to which this population was submitted and on the strategy it used to cope with them.

Take, for example, the distinction ecologists have proposed between *K* and *r* strategies. *K* and *r* refer to the parameters in logistic equations. Though this distinction is only relative, it is especially clear when it characterizes the divergence resulting from a systematic interaction between two populations, particularly the prey-predator interaction. In this view, the typical evolution for a prey population will be the increase in the reproduction rate *r*. The predator will evolve toward more effective ways of capturing its prey—that is, toward an amelioration of *K*. But this amelioration, defined in a logistic frame, is liable to have consequences that go beyond the situations defined by logistic equations.

As Stephen J. Gould remarked,[16] a *K* strategy implies individuals becoming more and more able to learn from experience and to store memories—that is, individuals more complex with a longer period of maturation and apprenticeship. This in turn means individuals both more "valuable"—representing a larger biological investment—and characterized by a longer period of vulnerability. The development of "social" and "family" ties thus appears as a logical counterpart of the *K* strategy. From that point on, other factors, besides the mere number of individuals in the population, become more and more relevant and the logistic equation measuring the success by the number of individuals becomes misleading. We have here a particular example of what makes modelization so risky. In complex systems, both the definition of entities and of the interactions among them can be modified by evolution. Not only each state of a system but also the very definition of the system as modelized is generally unstable, or at least metastable.

We come to problems where methodology cannot be separated from the question of the nature of the object investigated. We cannot ask the same questions about a population of flies that reproduce and die by millions without apparently learning from or enlarging their experience and about a population of primates where each individual is an entanglement of its own experiences and the traditions of the populations in which he lives.

We also find that, within anthropology itself, basic choices must be made between various approaches to collective phenomena. It is well known, for example, that structural anthropology privileges those aspects of society where the tools of logic and finite mathematics can be used, aspects such as the elementary structures of kinship or the analysis of myths, whose transformations are often compared to crystalline growth. Discrete elements are counted and combined. This contrasts with approaches that analyze evolution in terms of processes involving large, partially chaotic populations. We are dealing with two different outlooks and two types of models: Lévi-Strauss defines them respectively as "mechanical" and "statistical." In the mechanical model "the elements are of the same scale as the phenomena" and individual behavior is based on prescriptions referring to the structural organization of society. The anthropologist makes the logic of this behavior explicit. The sociologist, on the other hand, works with statistical models for large populations and defines averages and thresholds.[17]

A society defined entirely in terms of a functional model would correspond to the Aristotelian idea of natural hierarchy and order. Each official would perform the duties for which he has been appointed. These duties would translate at each level the different aspects of the organization of the society as a whole. The king gives orders to the architect, the architect to the contractor, the contractor to the workers. Everywhere a mastermind is at work. On the contrary, termites and other social insects seem to approach the "statistical" model. As we have seen, there seems to be no mastermind behind the construction of the termites' nest, when interactions among individuals produce certain types of collective behavior in some circumstances, but none of these interactions refer to any global task, being all purely local. Such a description necessarily implies averages and reintroduces the question of stability and bifurcations.

Which events will regress, and which are likely to affect the whole system? What are the situations of choice, and what are the regimes of stability? Since size or the system's density may play the role of a bifurcation parameter, how may purely quantitative growth lead to qualitatively new choices? Questions such as these call for an ambitious program indeed. As with the *r* and *K* strategies, they lead us to connect the choice of a "good" model for social behavior and history. How does the evolution of a population lead it to become more "mechanical"? This question seems parallel to questions we have already met in biology. How, for example, does the selection of the genetic information governing the rates and regulations of metabolic reactions favor certain paths to such an extent that development seems to be purposive or appear as the translation of a "message"?

We believe that models inspired by the concept of "order through fluctuations" will help us with these questions and even permit us in some circumstances to give a more precise formulation to the complex interplay between individual and collective aspects of behavior. From the physicist's point of view, this involves a distinction between states of the system in which all individual initiative is doomed to insignificance on the one hand, and on the other, bifurcation regions in which an individual, an idea, or a new behavior can upset the global state. Even in those regions, amplification obviously does not occur with just any individual, idea, or behavior, but only with those that are "dangerous"—that is, those that can exploit to their advantage the nonlinear relations guaranteeing the stability of the preceding regime. Thus we are led to conclude that *the same* nonlinearities may produce an order out of the chaos of elementary processes and still, under different circumstances, be responsible for the destruction of this same order, eventually producing a new coherence beyond another bifurcation.

"Order through fluctuations" models introduce an unstable world where small causes can have large effects, but this world is not arbitrary. On the contrary, the reasons for the amplification of a small event are a legitimate matter for rational inquiry. Fluctuations do not cause the transformation of a system's activity. Obviously, to use an image inspired by Maxwell, the match is responsible for the forest fire, but reference to a match does not suffice to understand the fire. Moreover, the fact that a fluctuation evades control does not mean that we cannot locate the reasons for the instability its amplification causes.

An Open World

In view of the complexity of the questions raised here, we can hardly avoid stating that the way in which biological and social evolution has traditionally been interpreted represents a particularly unfortunate use of the concepts and methods borrowed from physics[18]—unfortunate because the area of physics where these concepts and methods are valid was very restricted, and thus the analogies between them and social or economic phenomena are completely unjustified.

The foremost example of this is the paradigm of optimization. It is obvious that the management of human society as well as the action of selective pressures tends to optimize some aspects of behaviors or modes of connection, but to consider optimization as the key to understanding how populations and individuals survive is to risk confusing causes with effects.

Optimization models thus ignore both the possibility of radical transformations—that is, transformations that change the definition of a problem and thus the kind of solution sought—and the inertial constraints that may eventually force a system into a disastrous way of functioning. Like doctrines such as Adam Smith's invisible hand or other definitions of progress in terms of maximization or minimization criteria, this gives a reassuring representation of nature as an all-powerful and rational calculator, and of a coherent history characterized by global progress. To restore both inertia and the possibility of unanticipated events—that is, restore the open character of history—we must accept its fundamental uncertainty. Here we could use as a symbol the apparently accidental character of the great cretaceous extinction that cleared the path for the development of mammals, a small group of ratlike creatures.[19]

This has been a general presentation, a kind of "bird's-eye view," and thus has omitted many topics of great interest: flames, plasmas, and lasers, for example, present nonequilibrium instabilities of great theoretical and practical interest.

Everywhere we look, we find a nature that is rich in diversity and innovations. The conceptual evolution we have described is itself embedded in a wider history, that of the progressive rediscovery of time.

We have seen new aspects of time being progressively incorporated into physics, while the ambitions to omniscience inherent in classical science were progressively rejected. In this chapter we have moved from physics through biology and ecology to human society, but we could have proceeded in the inverse order. Indeed, history began by concentrating mainly on human societies, after which attention was given to the temporal dimensions of life and of geology. The incorporation of time into physics thus appears as the last stage of a progressive reinsertion of history into the natural and social sciences.

Curiously, at every stage of the process, a decisive feature of this "historicization" has been the discovery of some temporal heterogeneity. Since the Renaissance, Western society has come into contact with different populations that were seen as corresponding to different stages of development; nineteenth-century biology and geology learned to discover and classify fossils and to recognize in landscapes the memories of a past with which we coexist; finally, twentieth-century physics has also discovered a kind of fossil, residual black-body radiation, which tells us about the beginnings of the universe. Today we know that we live in a world where different interlocked times and the fossils of many pasts coexist.

We must now proceed to another question. We have said that life is starting to seem as "natural as a falling body." What has the natural process of self-organization to do with a falling body? What possible link can there be between dynamics, the science of force and trajectories, and the science of complexity and becoming, the science of living processes and of the natural evolution of which they are part? At the end of the nineteenth century, irreversibility was associated with the phenomena of friction, viscosity, and heating. Irreversibility lay at the origin of energy losses and waste. At that time it was still possible to subscribe to the fiction that irreversibility was only a result of our ineptitude, of our unsophisticated machines, and that nature remained fundamentally reversible. Now it is no longer possible: today even physics tells us that irreversible processes play a constructive and indispensable role.

So we come to a question that can be avoided no longer. What is the relation between this new science of complexity and the science of simple, elementary behavior? What is the relation between these two opposing views of nature? Are there two sciences, two truths for a single world? How is that possible?

In a certain sense, we have come back to the beginning of modern science. Now, as at Newton's time, two sciences come face to face—the science of gravitation, which describes an atemporal nature subject to laws, and the science of fire, chemistry. We now understand why it was impossible for the first synthesis produced by science, the Newtonian synthesis, to be complete; the forces of interaction described by dynamics cannot explain the complex and irreversible behavior of matter. *Ignis mutat res.* According to this ancient saying, chemical structures are the creatures of fire, the results of irreversible processes. How can we bridge the gap between being and becoming—two concepts in conflict, yet both necessary to reach a coherent description of this strange world in which we live?

BOOK THREE

FROM BEING TO BECOMING

CHAPTER VII

REDISCOVERING TIME

A Change of Emphasis

Whitehead wrote that a "clash of doctrines is not a disaster, it is an opportunity."[1] If this statement is true, few opportunities in the history of science have been so promising: two worlds have come face to face, the world of dynamics and the world of thermodynamics.

Newtonian science was the outcome, the crowning synthesis of centuries of experimentation as well as of converging lines of theoretical research. The same is true for thermodynamics. The growth of science is quite different from the uniform unfolding of scientific disciplines, each in turn divided into an increasing number of watertight compartments. Quite the contrary, the convergence of different problems and points of view may break open the compartments and stir up scientific culture. These turning points have consequences that go beyond their scientific context and influence the intellectual scene as a whole. Inversely, global problems often have been sources of inspiration to science.

The clash of doctrines, the conflict between being and becoming, indicates that a new turning point has been reached, that a new synthesis is needed. Such a synthesis is taking shape today, every bit as unexpected as the preceding ones. We again find a remarkable convergence of research, all of which contributes to identifying the difficulties inherent in the Newtonian concept of a scientific theory.

The ambition of Newtonian science was to present a vision of nature that would be universal, deterministic, and objective inasmuch as it contains no reference to the observer, complete inasmuch as it attains a level of description that escapes the clutches of time.

We have reached the core of the problem. "What is time?" Must we accept the opposition, traditional since Kant, between the static time of classical physics and the existential time we experience in our lives? According to Carnap:

> Once Einstein said that the problem of the Now worried him seriously. He explained that the experience of the Now means something special for man, something essentially different from the past and the future, but that this important difference does not and cannot occur within physics. That this experience cannot be grasped by science seemed to him a matter of painful but inevitable resignation. I remarked that all that occurs objectively can be described in science; on the one hand the temporal sequence of events is described in physics; and, on the other hand, the peculiarities of man's experiences with respect to time, including his different attitude towards past, present and future, can be described and (in principle) explained in psychology. But Einstein thought that these scientific descriptions cannot possibly satisfy our human needs; that there is something essential about the Now which is just outside of the realm of science.[2]

It is interesting to note that Bergson, in a sense following an opposite road, also reached a dualistic conclusion (see Chapter III). Like Einstein, Bergson started with a subjective time and then moved to time in nature, time as objectified by physics. However, for him this objectivization led to a debasement of time. Internal existential time has qualitative features that are lost in the process. It is for this reason that Bergson introduced the distinction between physical time and duration, a concept referring to existential time.

But we cannot stop here. As J. T. Fraser says, "The resulting dichotomy between time felt and time understood is a hallmark of scientific-industrial civilization, a sort of collective schizophrenia."[3] As we have already emphasized, where classical science used to emphasize permanence, we now find change and evolution; we no longer see in the skies the trajectories that filled Kant's heart with the same admiration as the moral law residing in him. We now see strange objects: quasars, pulsars, galaxies exploding and being torn apart, stars that, we are told, collapse into "black holes" irreversibly devouring everything they manage to ensnare.

Time has penetrated not only biology, geology, and the social sciences but also the two levels from which it has been traditionally excluded, the microscopic and the cosmic. Not only life, but also the universe as a whole has a history; this has profound implications.

The first theoretical paper dealing with a cosmological model from the point of view of general relativity was published by Einstein in 1917. It presented a static, timeless view of the universe, Spinoza's vision translated into physics. But then comes the unexpected. It became immediately evident that there were other, time-dependent solutions to Einstein's cosmological equations. We owe this discovery to the Russian astrophysicist A. Friedmann and the Belgian G. Lemaître. At the same time Hubble and his coworkers were studying the motions of galaxies, and they demonstrated that the velocity of distant galaxies is proportional to their distance from earth. The relation with the expanding universe discovered by Friedmann and Lemaître was obvious. Yet for many years physicists remained reluctant to accept such an "historical" description of cosmic evolution. Einstein himself was wary of it. Lemaître often said that when he tried to discuss with Einstein the possibility of making the initial state of the universe more precise and perhaps finding there the explanation of cosmic rays, Einstein showed no interest.

Today there is new evidence, the famous residual blackbody radiation, the light that illuminated the explosion of the hyperdense fireball with which our universe began. The whole story appears as another irony of history. In a sense, Einstein has, against his will, become the Darwin of physics. Darwin taught us that man is embedded in biological evolution; Einstein has taught us that we are embedded in an evolving universe. Einstein's ideas led him to a new continent, as unexpected to him as was America to Columbus. Einstein, like many physicists of his generation, was guided by a deep conviction that there was a fundamental, simple level in nature. Yet today this level is becoming less and less accessible to experiment. The only objects whose behavior is truly "simple" exist in our own world, at the macroscopic level. Classical science carefully chose its objects from this intermediate range. The first objects singled out by Newton—falling bodies, the pendulum, planetary motion—were simple. We know now, however, that this simplicity is not the hallmark of the fundamental: it cannot be attributed to the rest of the world.

Does this suffice? We now know that stability and simplicity are exceptions. Should we merely disregard the totalizing totalitarian claims of a conceptualization that, in fact, applies only to simple and stable objects? Why worry about the incompatibility between dynamics and thermodynamics?

We must not forget the words of Whitehead, words constantly confirmed by the history of science: a clash of doctrines is an opportunity, not a disaster. It has often been suggested that we simply ignore certain issues for *practical* reasons on the grounds that they are based on idealizations that are difficult to implement. At the beginning of this century, several physicists suggested abandoning determinism on the grounds that it was inaccessible in real experience.[4] Indeed, as we have already emphasized, we never know the exact positions and velocities of the molecules in a large system; thus an exact prediction of the system's future evolution is impossible. More recently, Brillouin hoped to destroy determinism by appealing to the commonsense truth that accurate prediction requires an accurate knowledge of the initial conditions and that this knowledge must be paid for; the exact prediction necessary to make determinism work requires that an "infinite" price be paid.

These objections, while reasonable, do not affect the conceptual world of dynamics. They shed no new light on reality. Moreover, the improvements in technology could bring us closer and closer to the idealization implied by classical dynamics.

In contrast, demonstrations of "impossibility" have a fundamental importance. They imply the discovery of an *unexpected intrinsic structure of reality* that dooms an intellectual enterprise to failure. Such discoveries will exclude the possibility of an operation that previously could have been imagined as feasible, at least in principle. "No engine can have an efficiency greater than one," "no heat engine can produce useful work unless it is in contact with two sources" are examples of statements of impossibility which have led to profound conceptual innovations.

Thermodynamics, relativity, and quantum mechanics are all rooted in the discovery of impossibilities, of limits to the ambitions of classical physics. Thus they marked the end of an exploration that had reached its limits. But we can now see these scientific innovations in a different light, not as an end but a beginning, as the opening up of new opportunities. We shall see in Chapter IX that the second law of thermodynamics expresses an "impossibility," even on the microscopic level, but even there the newly discovered impossibility becomes a starting point for the emergence of new concepts.

The End of Universality

Scientific description must be consistent with the resources available to an observer who belongs to the world he describes and cannot refer to some being who contemplates the physical world "from the outside." This is one of the fundamental requirements of relativity theory. In connection with the propagation of signals a limit appears that cannot be transgressed by any observer. Indeed, c, the velocity of light in vacuum ($c = 300{,}000$ km/sec), is the limiting velocity for the propagation of all signals. Thus this limiting velocity plays a fundamental role. It limits the region in space that may influence the point where an observer is located.

There is no universal constant in Newtonian physics. This is the reason for its claim to universality, why it can be applied in the same way whatever the scale of the objects: the motion of atoms, planets, and stars are governed by a single law.

The discovery of universal constants signified a radical change. Using the velocity of light as the comparison standard, physics has established a distinction between low and high velocities, those approaching the speed of light.

Likewise, Planck's constant, h, sets up a natural scale according to the object's mass. The atom can no longer be regarded as a tiny planetary system. Electrons belong to a different scale than planets and all other heavy, slow-moving, macroscopic objects, including ourselves.

Universal constants not only destroy the homogeneity of the universe by introducing physical scales in terms of which various behaviors become qualitatively different, they also lead to a new conception of objectivity. No observer can transmit signals at a velocity higher than that of light in a vacuum. Hence Einstein's remarkable conclusion: we can no longer define the absolute simultaneity of two distant events; simultaneity can be defined only in terms of a given reference frame. The scope of this book does not permit an extensive account of relativity theory. Let us merely point out that Newton's laws did not assume that the observer was a "physical being." Objective description was defined precisely as the absence of any reference to its author. For "nonphysical" intelligent beings capable of communicating at an infinite velocity, the laws of relativity would be irrelevant. The fact that relativity is based on a constraint that applies only to physically localized observers, to beings who can be in only one place at a time and not everywhere at once, gives this physics a "human" quality. This does not mean, however, that it is a "subjective" physics, the result of our preferences and convictions; it remains subject to intrinsic constraints that identify us as part of the physical world we are describing. It is a physics that presupposes an observer situated within the observed world. Our dialogue with nature will be successful only if it is carried on from within nature.

The Rise of Quantum Mechanics

Relativity altered the classical concept of objectivity. However, it left unchanged another fundamental characteristic of classical physics, namely, the ambition to achieve a "complete" description of nature. After relativity, physicists could no longer appeal to a demon who observed the entire universe from outside, but they could still conceive of a supreme mathematician who, as Einstein claimed, neither cheats nor plays dice. This mathematician would possess the formula of the universe, which would include a complete description of nature. In this sense, relativity remains a continuation of classical physics.

Quantum mechanics, on the other hand, is the first physical theory truly to have broken with the past. Quantum mechanics not only situates us in nature, it also labels us as "heavy" beings composed of a macroscopic number of atoms. In order to visualize more clearly the consequences of the velocity of light as a universal constant, Einstein imagined himself riding a photon. But quantum mechanics discovered that we are too heavy to ride photons or electrons. We cannot possibly replace such airy beings, identify ourselves with them, and describe what they would think, if they were able to think, and what they would experience, if they were able to feel anything.

The history of quantum mechanics, like that of all conceptual innovations, is complex, full of unexpected events; it is the history of a logic whose implications were discovered long after it was conceived in the urgency of experiment and in a difficult political and cultural environment.[5] This history cannot be related here; we only wish to emphasize its role in the construction of the bridge from being to becoming, which is our main subject.

The birth of quantum mechanics was in itself part of the quest for this bridge. Planck was interested in the interaction between matter and radiation. Underlying his work was the ambition to accomplish for the matter-light interaction what Boltzmann had achieved for the matter-matter interaction, namely, to discover a kinetic model for irreversible processes leading to equilibrium.[6] To his surprise, he was forced, in order to reach experimental results valid at thermal equilibrium, to assume that an exchange of energy between matter and radiation occurred only in discrete steps involving a new universal constant. This universal constant "h" measures the "size" of each step.

In this case, as in many others, the challenge of irreversibility led to decisive progress in physics.

This discovery remained isolated until Einstein presented the first general interpretation of Planck's constant. He understood that it had far-reaching implications for the nature of light. He introduced a revolutionary concept: the wave-particle duality of light.

Since the beginning of the nineteenth century, light had been associated with wave properties manifest in phenomena such as diffraction or interference. However, at the end of the nineteenth century, new phenomena were discovered, notably the photoelectric effect—that is, the expulsion of electrons as the result of the absorption of light. These new experimental results were difficult to explain in terms of the traditional wave properties of light. Einstein solved the riddle by assuming that light may be *both* wave and particle and that these two aspects are related through Planck's constant. More precisely, a light wave is characterized by its frequency ν and its wavelength λ; h permits us to go from frequency and wavelength to mechanical quantities such as energy ε and momentum p. The relations between ν and λ on the one side and ε and p on the other are very simple: $\varepsilon = h\nu$, $p = h/\lambda$, and both involve h. Twenty years later, Louis de Broglie extended this wave-particle duality from light to matter; thus the starting point for the modern formulation of quantum mechanics.

In 1913 Niels Bohr had linked the new quantum physics to the structure of atoms (and later of molecules). As a result of the wave-particle duality, he showed that there exist discrete sequences of electron orbits. When an atom is excited, the electron jumps from one orbit to another. At this very instant the atom emits or absorbs a photon the frequency of which corresponds to the difference between the energies characterizing the electron's motion in each of the two orbits. This difference is calculated in terms of Einstein's formula relating energy to frequency.

Thus we reach the decisive years 1925–27, a "golden age" of physics.[7] During this short period, Heisenberg, Born, Jordan, Schrödinger, and Dirac made quantum physics into a consistent new theory. This theory incorporates Einstein's and de Broglie's wave-particle duality in the framework of a new generalized form of dynamics: quantum mechanics. For our purposes here, the conceptual novelty of quantum mechanics is essential.

First and foremost, a new formulation, unknown in classical physics, had to be introduced to allow "quantitization" to be incorporated into the theoretical language. The essential fact is that an atom can be found only in discrete energy levels corresponding to the various electron orbits. In particular, this means that energy (or the Hamiltonian) can no longer be merely a function of the position and the moment, as it is in classical mechanics. Otherwise, by giving the positions and moments slightly different values, energy could be made to vary continuously. But as observation reveals, only discrete levels exist.

We therefore have to replace the conventional idea that the Hamiltonian is a function of position and momenta with something new; the basic idea of quantum mechanics is that the Hamiltonian as well as the other quantities of classical mechanics, such as coordinates q or momenta p, now become *operators*. This is one of the boldest ideas ever introduced in science, and we would like to discuss it in detail.

It is a simple idea, even if at first it seems somewhat abstract. We have to distinguish the operator—a mathematical operation—and the object on which it operates—a function. As an example, take as the mathematical "operator" the derivative represented by d/dx and suppose it acts on a function—say, x^2; the result of this operation is a *new* function, this time "$2x$." However, certain functions behave in a peculiar way with respect to derivation. For example, the derivative of "e^{3x}" is "$3e^{3x}$": here we return to the original function simply multiplied by some number—here, 3. Functions that are merely recovered by a given operator to them are known as the "eigenfunctions" of this operator, and the numbers by which the eigenfunction is multiplied after the application of the operator are the "eigenvalues" of the operator.

To each operator there thus corresponds an ensemble, a "reservoir" of numerical values; this ensemble forms its "spectrum." This spectrum is "discrete" when the eigenvalues form a discrete series. There exists, for instance, an operator with all the integers 0, 1, 2 . . . as eigenvalues. A spectrum may also be continuous—for example, when it consists of all the numbers between 0 and 1.

The basic concept of quantum mechanics may thus be expressed as follows: to all physical quantities in classical mechanics there corresponds in quantum mechanics an operator, and the numerical values that may be taken by this physical quantity are the eigenvalues of this operator. The essential point is that the concept of physical quantity (represented by an operator) is now distinct from that of its numerical values (represented by the eigenvalues of the operator). In particular, energy will now be represented by the Hamiltonian operator, and the energy levels—the observed values of the energy—will be identified with the eigenvalues corresponding to this operator.

The introduction of operators opened up to physics a microscopic world of unsuspected richness, and we regret that we cannot devote more space to this fascinating subject, in which creative imagination and experimental observation are so successfully combined. Here we wish merely to stress that the microscopic world is governed by laws having a new structure, thereby putting an end once and for all to the hope of discovering a single conceptual scheme common to all levels of description.

A new mathematical language invented to deal with a certain situation may actually open up fields of inquiry that are full of surprises, going far beyond the expectations of its originators. This was true for differential calculus, which lies at the root of the formulation of classical dynamics. It is true as well for operator calculus. Quantum theory, initiated as demanded by the result of unexpected experimental discoveries, was quick to reveal itself as pregnant with new content.

Today, more than fifty years after the introduction of operators into quantum mechanics, their significance remains a subject of lively discussion. From the historical point of view, the introduction of operators is linked to the existence of energy levels, but today operators have applications even in classical physics. This implies that their significance has been extended beyond the expectations of the founders of quantum mechanics. Operators now come into play as soon as, for one reason or another, the notion of a dynamic trajectory has to be discarded, and with it, the deterministic description a trajectory implies.

Heisenberg's Uncertainty Relation

We have seen that in quantum mechanics to each physical quantity corresponds an operator that acts on functions. Of special importance are the eigenfunctions and the eigenvalues corresponding to the operator under consideration. The eigenvalues correspond precisely to the numerical values the physical quantity can now take. Let us take a closer look at the operators quantum mechanics associates with coordinates q and momenta p; their coordinates are, as we have seen in Chapter II, the canonical variables.

In classical mechanics coordinates and momenta are independent in the sense that we can ascribe to a coordinate a numerical value quite independent of the value we have ascribed to the momentum. However, the existence of Planck's constant h implies the reduction in the number of independent variables. We could have guessed this right away from the Einstein-de Broglie relation $\lambda = h/p$, which, as we have seen, connects wavelength to momentum. Planck's constant h expresses a relation between lengths (closely related to the concept of coordinates) and momenta. Therefore, positions and momenta can no longer be independent variables, as in classical mechanics. The operators corresponding to positions and momenta can be expressed in terms of the coordinate alone or in terms of the momentum, something explained in all textbooks dealing with quantum mechanics.

The important point is that in all cases, only one type of quantity appears (either coordinate or momentum), but not both. In this sense we may say that the quantum mechanics divides the number of classical mechanical variables by a factor of two.

One fundamental property results from the relation between operators in quantum mechanics: the two operators q_{op} and p_{op} do not *commute*—that is, the results of $q_{op}p_{op}$ and of $p_{op}q_{op}$ applied to the same function are different. This has profound implications, since only commuting operators admit common eigenfunctions. Thus we cannot identify a function that would be an eigenfunction of both coordinate and momentum. As a consequence of the definition of the coordinate and momentum operators in quantum mechanics, there can be no state in which the physical quantities, coordinate q and momentum p, both have a well-defined value. This situation, unknown in classical mechanics, is expressed by Heisenberg's famous uncertainty relations. We can measure a coordinate and a momentum, but the dispersions of the respective possible predictions as expressed by $\Delta q, \Delta p$ are related by the Heisenberg inequality $\Delta q \Delta p \geq h$. We can make Δq as small as we want, but then Δp goes to infinity, and vice versa.

Much has been written about Heisenberg's uncertainty relations, and our discussion is admittedly oversimplified. But we wish to give our readers some understanding of the new problem that results from the use of operators: Heisenberg's uncertainty relation necessarily leads to a revision of the concept of causality. It is possible to determine the coordinate precisely. But the moment we do so, the momentum will acquire an arbitrary value, positive or negative. In other words, in an instant the position of the object will become arbitrarily distant. The meaning of localization becomes blurred: the concepts that form the basis of classical mechanics are profoundly altered.

These consequences of quantum mechanics were unacceptable to many physicists, including Einstein; and many experiments were devised to demonstrate their absurdity. An attempt was also made to minimize the conceptual change involved. In particular, it was suggested that the foundation of quantum mechanics is in some way related to perturbations resulting from the process of observation. A system was thought to possess intrinsically well-defined mechanical parameters such as coordinates and momenta; but some of them would be made fuzzy by measurement, and Heisenberg's uncertainty relation would only express the perturbation created by the measurement process. Classical realism thus would remain intact on the fundamental level, and we would simply have to add a positivistic qualification. This interpretation seems too narrow. It is not the quantum measurement process that disturbs the results. Far from it: Planck's constant forces us to revise our concepts of coordinates and momenta. This conclusion has been confirmed by recent experiments designed to test the assumption of local hidden variables that were introduced to restore classical determinism.[8] The results of those experiments confirm the striking consequences of quantum mechanics.

That quantum mechanics obliges us to speak less absolutely about the localization of an object implies, as Niels Bohr often emphasized, that we must give up the realism of classical physics. For Bohr, Planck's constant defines the interaction between a quantum system and the measurement device as nondecomposable. It is only to the quantum phenomenon as a whole, including the measurement interaction, that we can ascribe numerical values. All description thus implies a choice of the measurement device, a choice of the question asked. In this sense, the answer, the result of the measurement, does not give us access to a given reality. We have to decide which measurement we are going to perform and which question our experiments will ask the system. Thus there is an irreducible multiplicity of representations for a system, each connected with a determined set of operators.

This implies a departure from the classical notion of objectivity, since in the classical view the only "objective" description is the complete description of *the system as it is,* independent of the choice of how it is observed.

Bohr always emphasized the novelty of the *positive* choice introduced through measurement. The physicist has to choose his language, to choose the macroscopic experimental device. Bohr expressed this idea through the principle of complementarity,[9] which may be considered as an extension of Heisenberg's uncertainty relations. We can measure coordinates or momenta, but not both. No single theoretical language articulating the variables to which a well-defined value can be attributed can exhaust the physical content of a system. Various possible languages and points of view about the system may be complementary. They all deal with the same reality, but it is impossible to reduce them to one single description. The irreducible plurality of perspectives on the same reality expresses the impossibility of a divine point of view from which the whole of reality is visible. However, the lesson of the principle of complementarity is not a lesson in resignation. Bohr used to say that the significance of quantum mechanics always made him dizzy, and we do indeed feel dizzy when we are torn from the comfortable routine of common sense.

The real lesson to be learned from the principle of complementarity, a lesson that can perhaps be transferred to other fields of knowledge, consists in emphasizing the wealth of reality, which overflows any single language, any single logical structure. Each language can express only part of reality. Music, for example, has not been exhausted by any of its realizations, by any style of composition, from Bach to Schönberg.

We have emphasized the importance of operators because they demonstrate that the reality studied by physics is also a mental construct; it is not merely given. We must distinguish between the abstract notion of a coordinate or of momentum, represented mathematically by operators, and their numerical realization, which can be reached through experiments. One of the reasons for the opposition between the "two cultures" may have been the belief that literature corresponds to a conceptualization of reality, to "fiction," while science seems to express objective "reality." Quantum mechanics teaches us that the situation is not so simple. On all levels reality implies an essential element of conceptualization.

The Temporal Evolution of Quantum Systems

We shall now move on to discuss the temporal evolution of quantum systems. As in classical mechanics, the Hamiltonian plays a fundamental role. As we have seen, in quantum mechanics it is replaced by the Hamiltonian operator H_{op}. This energy operator plays a central role: on the one hand, its eigenvalues correspond to the energy levels; on the other hand, as in classical mechanics, the Hamiltonian operator determines the temporal evolution of the system. In quantum mechanics the role played by the canonical equation of classical mechanics is taken by the Schrödinger equation, which expresses the time evolution of the function characterizing the quantum state as the result of the application of the operator H_{op} on the wave function ψ (there are, of course, other formulations, which we cannot describe here). The term "wave function" has been chosen to emphasize once again the wave-particle duality so fundamental in all of quantum physics. ψ *is a wave amplitude* that evolves according to a particle type of equation determined by the Hamiltonian. Schrödinger's equation, like the canonical equation of classical physics, expresses a *reversible* and *deterministic* evolution. The reversible change of wave function corresponds to a reversible motion along a trajectory. If the wave function at a given instant is known, Schrödinger's equation allows it to be calculated for any previous or subsequent instant. From this viewpoint, the situation is strictly similar to that in classical mechanics. This is because the uncertainty relations of quantum mechanics do not include time. Time remains a number, not an operator, and only operators can appear in Heisenberg's uncertainty relations.

Quantum mechanics deals with only half of the variables of classical mechanics. As a result, classical determinism becomes inapplicable, and in quantum physics statistical considerations play a central role. It is through the wave intensity $|\psi^2|$ (the square of the amplitude) that we make contact with statistical considerations.

The standard statistical interpretation of quantum mechanics runs as follows: consider the eigenfunctions of some operator—say, the energy operator H_{op}—and the corresponding eigenvalues. In general the wave function ψ will not be the eigenfunction of the energy operator, but it can be expressed as the superposition of these eigenfunctions. The respective importance of each eigenfunction in this superposition allows us to calculate the probability for the appearance of the various possible corresponding eigenvalues.

Here again we notice a fundamental departure from classical theory. Only probabilities can be predicted, not single events. This was the second time in the history of science that probabilities were used to explain some basic features of nature. The first time was in Boltzmann's interpretation of entropy. There, however, a subjective point of view remained possible; in this view, "only" our ignorance in the face of the complexity of the systems considered prevented us from achieving a complete description. (We shall see that today it is possible to overcome this attitude.) Here, as before, the use of probabilities was unacceptable to many physicists—including Einstein—who wished to achieve a "complete" deterministic description. Just as with irreversibility, an appeal to our ignorance seemed to offer a way out: our inaptitude would make us responsible for statistical behavior in the quantum world, just as it makes us responsible for irreversibility.

Once again we come to the problem of hidden variables. However, as we have said, there has been no experimental evidence to justify the introduction of such variables, and the role of probabilities seems irreducible.

There is only one case in which the Schrödinger equation leads to a deterministic prediction: that is when ψ, instead of being a *superposition* of eigenfunctions, is *reduced* to a single one. In particular, in an ideal measurement process, a system may be prepared in such a way that the result of a given measurement may be predicted. We then know that the system is described by the corresponding eigenfunction. From then on, the system may be described with certainty as being in the eigenstate indicated by the measurement result.

The measurement process in quantum mechanics has a special significance that is attracting considerable interest today. Suppose we start with a wave function, which is indeed a superposition of eigenfunctions. As a result of the measurement process, this single collection of systems all represented by the same wave function is replaced by a collection of wave functions corresponding to the various eigenvalues that may be measured. Stated technically, a measurement leads from a single wave function (a "pure" state) to a mixture.

As Bohr and Rosenfeld[10] repeatedly pointed out, every measurement contains an element of *irreversibility,* an appeal made to irreversible phenomena, such as chemical processes corresponding to the recording of the "data." Recording is accompanied by an amplification whereby a microscopic event produces an effect on a macroscopic level—that is, a level at which we can read the measuring instruments. The measurement thus presupposes irreversibility.

This was in a sense already true in classical physics. However, the problem of the irreversible character of measurement is more urgent in quantum mechanics because it raises questions at the level of its formulation.

The usual approach to this problem states that quantum mechanics has no choice but to postulate the coexistence of two mutually irreducible processes, the reversible and continuous evolution described by Schrödinger's equation and the irreversible and discontinuous reduction of the wave function to one of its eigenfunctions at the time of measurement. Thus the paradox: the reversible Schrödinger equation can be tested only by irreversible measurements that the equation is *by definition* unable to describe. It is thus impossible for quantum mechanics to set up a closed structure.

In the face of these difficulties, some physicists have once more taken refuge in subjectivism, stating that *we—our* measurement and even, for some, *our* mind—determine the evolution of the system that breaks the law of natural, "objective" reversibility.[11] Others have concluded that Schrödinger's equation was not "complete" and that new terms must be added to account for the irreversibility of the measurement. Other more improbable "solutions" have also been proposed, such as Everett's many-world hypothesis (see d'Espagnat, ref. 8). For us, however, the coexistence in quantum mechanics of reversibility and irreversibility shows that the classical idealization that describes the dynamic world as self-contained is impossible at the microscopic level. This is what Bohr meant when he noted that the language we use to describe a quantum system cannot be separated from the macroscopic concepts that describe the functioning of our measurement instruments. Schrödinger's equation does not describe a separate level of reality; rather it presupposes the macroscopic world to which we belong.

The problem of measurement in quantum mechanics is thus an aspect of one of the problems to which this book is devoted—the connection between the simple world described by Hamiltonian trajectories and Schrödinger's equation, and the complex macroscopic world of irreversible processes.

In Chapter IX, we shall see that irreversibility enters classical physics when the idealization involved in the concept of a trajectory becomes inadequate. The measurement problem in quantum mechanics is susceptible to the same type of solution.[12] Indeed, the wave function represents the maximum knowledge of a quantum system. As in classical physics, the object of this maximum knowledge satisfies a reversible evolution equation. In both cases, irreversibility enters when the ideal object corresponding to maximum knowledge has to be replaced by less idealized concepts. But when does this happen? This is the question of the physical mechanisms of irreversibility to which we shall turn in Chapter IX. But let us first summarize some other features of the renewal of contemporary science.

A Nonequilibrium Universe

The two scientific revolutions described in this chapter started as attempts to incorporate universal constants, c and h, into the framework of classical mechanics. This led to far-reaching consequences, some of which we have described here. From other perspectives, relativity and quantum mechanics seemed to adhere to the basic world view expressed in Newtonian mechanics. This is especially true regarding the role and meaning of time. In quantum mechanics, once the wave function at time zero is known, its value $\psi(t)$ both for future and past is determined. Likewise, in relativity theory the *static* geometric character of time is often emphasized by the use of four-dimensional notation (three dimensions for space and one for time). As expressed concisely by Minkowski in 1908, "space by itself and time by itself are doomed to fade away into mere shadows, and only a kind of union of the two will preserve an independent reality . . . only a world in itself will subsist."[13]

But over the past five decades this situation has radically changed. Quantum mechanics has become the main tool for dealing with elementary particles and their transformations. It is outside the scope of this book to describe the bewildering variety of elementary particles that have appeared during the past few years.

We want only to recall that, using both quantum mechanics and relativity, Dirac demonstrated that we have to associate to each particle of mass m and charge e an antiparticle of the same mass but of opposite charge. Positrons, the antiparticles of electrons, as well as antiprotons, are currently being produced in high-energy accelerators. Antimatter has become a common subject of study in particle physics. Particles and their corresponding antiparticles annihilate each other when they collide, producing photons, massless particles corresponding to light. The equations of quantum theory are symmetric in respect to the exchange particle-antiparticle, or more precisely, they are symmetric in respect to a weaker requirement known as the CPT symmetry. In spite of this symmetry, there exists a remarkable *dissymmetry* between particles and antiparticles in the world around us. We are made of particles (electrons, protons), while antiparticles remain rare laboratory products. If particles and antiparticles coexisted in equal amount, all matter would be annihilated. There is strong evidence that antimatter does not exist in our galaxy, but the possibility that it exists in distant galaxies cannot be excluded. We can imagine a mechanism in the universe that separates particles and antiparticles, hides antiparticles somewhere. However, it seems more likely that we live in a "nonsymmetrical" universe where matter completely dominates antimatter.

How is this possible? A model explaining the situation was presented by Sakharov in 1966, and today much work is being done along these lines.[14] One essential element of the model is that, at the time of the formation of matter, the universe had to be in *nonequilibrium conditions,* for at equilibrium the law of mass action discussed in Chapter V would have required equal amounts of matter and antimatter.

What we want to emphasize here is that nonequilibrium has now acquired a new, cosmological dimension. Without nonequilibrium and without the irreversible processes linked to it, the universe would have a completely different structure. There would be no appreciable amount of matter, only some fluctuating local excesses of matter over antimatter, or vice versa.

From a mechanistic theory that was modified to account for the existence of the universal constant h, quantum theory has evolved into a theory of mutual transformations of elementary particles. In recent attempts to formulate a "unified theory of elementary particles" it has even been suggested that *all* particles of matter, including the proton, are unstable (however, the lifetime of the proton would be enormous, of the order of 10^{30} years). Mechanics, the science of motion, instead of corresponding to the fundamental level of description, becomes a mere approximation, useful only because of the long lifetime of elementary particles such as protons.

Relativity theory has gone through the same transformations. As we mentioned, it started as a geometric theory that strongly emphasized timeless features. Today it is the main tool for investigating the thermal history of the universe, for providing clues to the mechanisms that led to the present structure of the universe. The problem of time, of irreversibility, has therefore acquired a new urgency. From the field of engineering, of applied chemistry, where it was first formulated, it has spread to the whole of physics, from elementary particles to cosmology.

From the perspective of this book, the importance of quantum mechanics lies in its introduction of probability into microscopic physics. This should not be confused with the stochastic processes that describe chemical reactions as discussed in Chapter V. In quantum mechanics, the wave function evolves in a deterministic fashion, except in the measurement process.

We have seen that in the fifty years since the formulation of quantum mechanics the study of nonequilibrium processes has revealed that fluctuations, stochastic elements, are important even on the microscopic scale. We have repeatedly stated in this book that the reconceptualization of physics going on today leads from deterministic, reversible processes to stochastic and irreversible ones. We believe that quantum mechanics occupies a kind of intermediate position in this process. There probability appears, but not irreversibility. We expect, and we shall give some reasons for this in Chapter IX, that the next step will be the introduction of fundamental irreversibility on the microscopic level. In contrast with the attempts to restore classical orthodoxy through hidden variables or other means, we shall argue that it is necessary to move even farther away from deterministic descriptions of nature and adopt a statistical, stochastic description.

CHAPTER VIII

THE CLASH OF DOCTRINES

Probability and Irreversibility

> We shall see that nearly everywhere the physicist has purged from his science the use of one-way time, as though aware that this idea introduces an anthropomorphic element, alien to the ideals of physics. Nevertheless, in several important cases unidirectional time and unidirectional causality have been invoked, but always, as we shall proceed to show, in support of some false doctrine.
>
> G. N. LEWIS[1]

> The law that entropy always increases—the second law of thermodynamics—holds, I think, the supreme position among the laws of Nature. If someone points out to you that your pet theory of the universe is in disagreement with Maxwell's equations—then so much the worse for Maxwell's equations. If it is found to be contradicted by observation—well, these experimentalists do bungle things sometimes. But if your theory is found to be against the second law of thermodynamics I can give you no hope; there is nothing for it but to collapse in deepest humiliation.
>
> A. S. EDDINGTON[2]

With Clausius' formulation of the second law of thermodynamics, the conflict between thermodynamics and dynamics became obvious. There is hardly a single question in physics that has been more often and more actively discussed than the relation between thermodynamics and dynamics. Even now, a hundred and fifty years after Clausius, the question still arouses strong feelings. No one can remain neutral in this conflict, which involves the meaning of reality and time. Must dynamics, the mother of modern science, be abandoned in favor of some form of thermodynamics? That was the view of the "energeticists," who exerted great influence during the nineteenth century. Is there a way to "save" dynamics, to recoup the second law without giving up the formidable structure built by Newton and his successors? What role can entropy play in a world described by dynamics?

We have already mentioned the answer proposed by Boltzmann. Boltzmann's famous equation $S=k \log P$ relates entropy and probability: entropy grows because probability grows. Let us immediately emphasize that in this perspective the second law would have great practical importance but would be of no fundamental significance. In his excellent book *The Ambidextrous Universe,* Martin Gardner writes: "Certain events go only one way not because they can't go the other way but because it is extremely unlikely that they go backward."[3] By improving our abilities to measure more and more unlikely events, we could reach a situation in which the second law would play as small a role as we want. This is the point of view that is often taken today. However, this was not Planck's point of view:

> It would be absurd to assume that the validity of the second law depends in any way on the skill of the physicist or chemist in observing or experimenting. The gist of the second law has nothing to do with experiment; the law asserts briefly that *there exists in nature a quantity which changes always in the same sense in all natural processes*. The proposition stated in this general form may be correct or incorrect; but whichever it may be, it will remain so, irrespective of whether thinking and measuring beings exist on the earth or not, and whether or not, assuming they do exist, they are able to measure the details of physical or chemical processes more accurately by one, two, or a hundred decimal places than we can. The limitation to the law, if any, must lie in the same province as its essential idea, in the observed Nature, and

> not in the Observer. That man's experience is called upon in the deduction of the law is of no consequence; for that is, in fact, our only way of arriving at a knowledge of natural law.[4]

However, Planck's views remained isolated. As we noted, most scientists considered the second law the result of approximations, the intrusion of subjective views into the exact world of physics. For example, in a celebrated sentence Born stated, "Irreversibility is the effect of the introduction of ignorance into the basic laws of physics."[5]

In the present chapter we wish to describe some of the basic steps in the development of the interpretation of the second law. We must first understand why this problem appeared to be so difficult. In Chapter IX we shall go on to present a new approach that, we hope, will clearly express both the radical originality and the objective meaning of the second law. Our conclusion will agree with Planck's view. We shall show that, far from destroying the formidable structure of dynamics, the second law adds an essential new element to it.

First we wish to clarify Boltzmann's association of probability and entropy. We shall begin by describing the "urn model" proposed by P. and T. Ehrenfest.[6] Consider N objects (for example, balls) distributed between two containers A and B. At regular time intervals (for example, every second) a ball

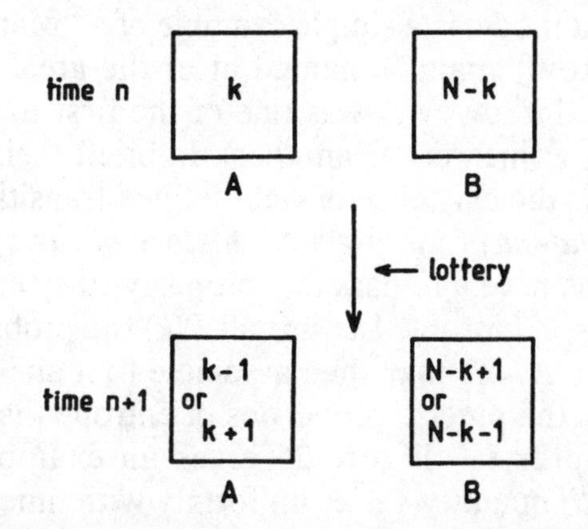

Figure 23. Ehrenfest's urn model. N balls are distributed between two containers A and B. At time n there are k balls in A and $N-k$ balls in B. At regular time intervals a ball is taken at random from one container and put in the other.

is chosen at random and moved from one container to the other. Suppose that at time n there are k balls in A and $N-k$ balls in B. Then at time $n+1$ there can be in A either $k-1$ or $k+1$ balls. We have the transition probabilities k/N for $k \rightarrow k-1$ and $1-k/N$ for $k \rightarrow k+1$. Suppose we continue the game. We expect that as a result of the exchanges of balls the most probable distribution in Boltzmann's sense will be reached. When the number N of balls is large, this distribution corresponds to an equal number $N/2$ of balls in each urn. This can be verified by elementary calculations or by performing the experiment.

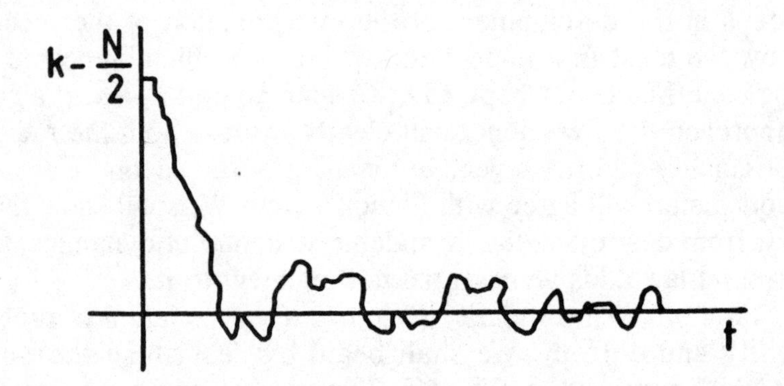

Figure 24. Approach to equilibrium ($k=N/2$) in Ehrenfest's urn model (schematic representation).

The Ehrenfest model is a simple example of a "Markov process" (or Markov "chain"), named after the great Russian mathematician Markov, who was one of the first to describe such processes (Poincaré was another). In brief, their characteristic feature is the existence of well-defined transition probabilities *independent of the previous history of the system.*

Markov chains have a remarkable property: they can be described in terms of entropy. Let us call $P(k)$ the probability of finding k balls in A. We may then associate to it an "$\mathcal{H}$ quantity," which has the precise properties of entropy that we discussed in Chapter IV. Figure 25 gives an example of its evolution. The $\mathcal{H}$ quantity varies uniformly with time, as does the entropy of an isolated system. It is true that $\mathcal{H}$ *decreases* with time, while the entropy S increases, but that is a matter of definition: $\mathcal{H}$ plays the role of $-S$.

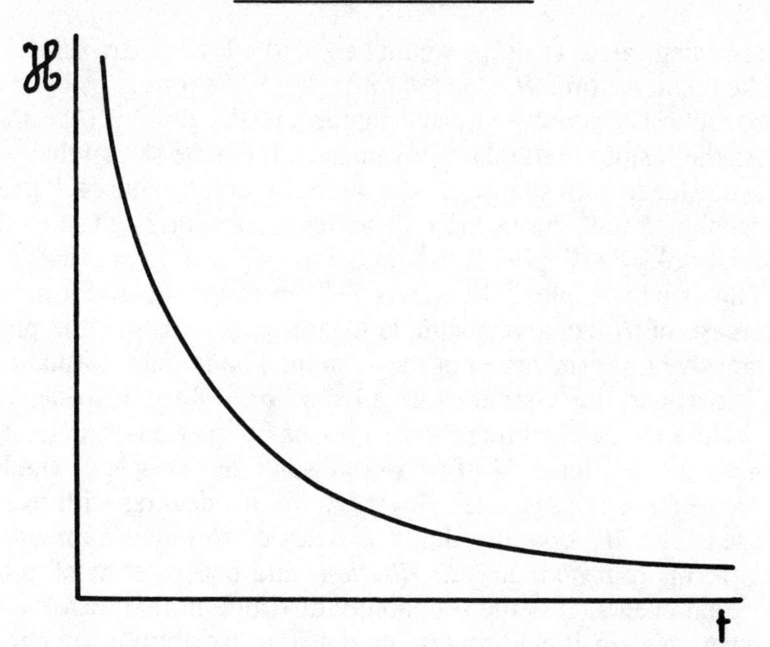

Figure 25. Time evolution of the $\mathcal{H}$ quantity (defined in the text) corresponding to the Ehrenfest model. This quantity decreases monotonously and vanishes for long times.

The mathematical meaning of this "$\mathcal{H}$ quantity" is worth considering in more detail: it measures the difference between the probabilities at a given time and those that exist at the equilibrium state (where the number of balls in each urn is $N/2$). The argument used in the Ehrenfest urn model can be generalized. Let us consider the partition of a square—that is, we subdivide the square into a number of disjointed regions. Then we consider the distribution of particles in the square and call $P(k,t)$ the probability of finding a particle in the region k. Similarly, we call $P_{eqm}(k)$ this quantity when uniformity is reached. We assume that, as in the urn model, there exist well-defined transition probabilities. The definition of the $\mathcal{H}$ quantity is

$$\mathcal{H} = \sum_{k} P(k,t) \, log \, \frac{P(k,t)}{P_{eqm}(k)}$$

Note the ratio $P(k,t)/P_{eqm}(k)$ that appears in this formula. Suppose there are eight boxes and that $P_{eqm}(k) = 1/8$. For example, we may start with all the particles in the first box. The corresponding values of $P(k,t)$ would be $P(1,t)=1$, all others zero. As the result we find $\mathcal{H}=\log(1/[1/8])=\log 8$. As time goes by, the particles become equally distributed and $P(k,t)=P_{\mathrm{eqm}}(k)=1/8$. As the result the $\mathcal{H}$ quantity vanishes. It can be shown that, in accordance with Figure 25, the decrease in the value of $\mathcal{H}$ proceeds in a uniform fashion. (The demonstration is given in all textbooks dealing with the theory of stochastic processes.) This is why $\mathcal{H}$ plays the role of $-S$, entropy. The uniform decrease of $\mathcal{H}$ has a very simple meaning: it measures the progressive uniformization of the system. The initial information is lost, and the system evolves from "order" to "disorder."

Note that a Markov process implies fluctuations, as clearly indicated in Figure 24. If we would wait long enough we would recover the initial state. However, we are dealing with averages. The $\mathcal{H}_M$ quantity that decreases uniformly is expressed in terms of *probability distributions* and not in terms of individual events. It is the probability distribution that evolves irreversibly (in the Ehrenfest model, the distribution function tends uniformly to a binomial distribution). Therefore, on the level of distribution functions, Markov chains lead to a *one-wayness in time*.

This arrow of time marks the difference between Markov chains and temporal evolution in quantum mechanics, where the wave function, though related to probabilities, evolves *reversibly*. It also illustrates the close relation between stochastic processes, such as Markov chains, and irreversibility. However, the increasing of entropy (or decreasing of $\mathcal{H}$) is not based on an arrow of time present in the laws of nature but on our decision to use present knowledge to predict future (and not past) behavior. Gibbs states it in his usual lapidary manner:

> But while the distinction of prior and subsequent events may be immaterial with respect to mathematical fictions, it is quite otherwise with respect to the events of the real world. It should not be forgotten, when our ensembles are chosen to illustrate the probabilities of events in the real world, that while the probabilities of subsequent events may often be determined from the probabilities of prior events, it is rarely the case that probabilities of prior

> events can be determined from those of subsequent events, for we are rarely justified in excluding the consideration of the antecedent probability of the prior events.[7]

It is an important point, which has led to a great deal of discussion.[8] Probability calculus is indeed time-oriented. The prediction of the future is different from retrodiction. If this was the whole story, we would have to conclude that we are forced to accept a subjective interpretation of irreversibility, since the distinction between future and past would depend only on us. In other words, in the subjective interpretation of irreversibility (further reinforced by the ambiguous analogy with information theory), the observer is responsible for the temporal asymmetry characterizing the system's development. Since the observer cannot in a single glance determine the positions and velocities of all the particles composing a complex system, he cannot know the instantaneous state that simultaneously contains its past and its future, nor can he grasp the reversible law that would allow him to predict its developments from one moment to the next. Neither can he manipulate the system like the demon invented by Maxwell, who can separate fast- and slow-moving particles and impose on a system an antithermodynamic evolution toward an increasingly less uniform temperature distribution.[9]

Thermodynamics remains the science of complex systems; but, from this perspective, the only specific feature of complex systems is that our knowledge of them is limited and that our uncertainty increases with time. Instead of recognizing in irreversiblity something that links nature to the observer, the scientist is compelled to admit that nature merely mirrors his ignorance. Nature is silent; irreversibility, far from rooting us in the physical world, is merely the echo of human endeavor and of its limits.

However, one immediate objection can be raised. According to such interpretations, thermodynamics ought to be as universal as our ignorance. There should exist *only* irreversible processes. This is the stumbling block for all universal interpretations of entropy that concentrate on our ignorance of initial (or boundary) conditions. *Irreversibility is not a universal property.* In order to link dynamics and thermodynamics, a physical criterion is required to distinguish between reversible and irreversible processes.

We shall take up this question in Chapter IX. Here let us return to the history of science and Boltzmann's pioneering work.

Boltzmann's Breakthrough

Boltzmann's fundamental contribution dates from 1872, about thirty years before the discovery of Markov chains. His ambition was to derive a "mechanical" interpretation of entropy. In other words, while in Markov chains the transition probabilities are given from outside as, for example, in the Ehrenfest model, we now have to relate them to the dynamic behavior of the system. Boltzmann was so fascinated by this problem that he devoted most of his scientific life to it. In his *Populäre Schriften*[10] he wrote: "If someone asked me what name we should give to this century, I would answer without hesitation that this is the century of Darwin." Boltzmann was deeply attracted by the idea of evolution, and his ambition was to become the "Darwin" of the evolution of matter.

The first step toward the mechanistic interpretation of entropy was to reintroduce the concept of "collisions" of molecules or atoms into the physical description, and along with it the possibility of a statistical description. This step had been taken by Clausius and Maxwell. Since collisions are discrete events, we may count them and estimate their average frequency. We may also classify collisions—for example, distinguish between collisions producing a particle with a given velocity v and collisions destroying a particle with a velocity v, producing molecules with a different velocity (the "direct" and "inverse" collisions).[11]

The question Maxwell asked was whether it was possible to define a state of a gas such that the collisions that incessantly modify the velocities of the molecules no longer determine any evolution in the *distribution* of these velocities—that is, in the mean number of particles for each velocity value. What is the velocity distribution such that the effects of the different collisions compensate each other on the population scale?

Maxwell demonstrated that this particular state, which is the thermodynamic equilibrium state, occurs when the velocity distribution becomes the well-known "bell-shaped curve," the "gaussian," which Quetelet, the founder of "social physics," had considered to be the very expression of randomness. Maxwell's theory permits us to give a simple interpretation of some of the basic laws describing the behavior of gases. An increase in temperature corresponds to an increase in the mean velocity of the molecules and thus of the energy associated with their motion. Experiments have verified Maxwell's law with great accuracy, and it still provides a basis for the solution of numerous problems in physical chemistry (for example, the calculation of the number of collisions in a reactive mixture).

Boltzmann, however, wanted to go farther. He wanted to describe not only the *state* of equilibrium but also *evolution* toward equilibrium—that is, evolution toward the Maxwellian distribution. He wanted to discover the molecular mechanism that corresponds to the increase of entropy, the mechanism that drives a system from an arbitrary distribution of velocities toward equilibrium.

Characteristically, Boltzmann approached the question of physical evolution not at the level of individual trajectories but at the level of a *population* of molecules. This, Boltzmann felt, was virtually tantamount to accomplishing Darwin's feat, but this time in physics: the driving force behind biological evolution—natural selection—cannot be defined for one individual but only for a large population. It is therefore a statistical concept.

Boltzmann's result may be described in relatively simple terms. The evolution of the distribution function $f(v, t)$ of the velocities v in some region of space and at time t appears as the sum of two effects; the number of particles at any given time t having a velocity v varies both as the result of the free motion of the particles and as the result of collisions between particles. The first result can be easily calculated in the terms of classical dynamics. It is in the investigation of the second result, due to collisions, that the originality of Boltzmann's method lies. In the face of the difficulties involved in following the trajectories (including the interactions), Boltzmann came to use concepts similar to those outlined in Chapter V (in connection with chemical reactions) and to calculate the *average* number of collisions creating or destroying a molecule corresponding to a velocity v.

Here once again there are two processes with opposite effects—"direct" collisions, those producing a molecule with velocity v starting from two molecules with velocities v' and v'', and "inverse" collisions, in which a molecule with velocity v is destroyed by collision with a molecule with velocity v'''. As with chemical reactions (see Chapter V, section 1), the frequency of such events is evaluated as being proportional to the product of the number of molecules taking part in these processes. (Of course, historically speaking, Boltzmann's method [1872] preceded that of chemical kinetics.)

The results obtained by Boltzmann are quite similar to those obtained in Markov chains. Again we shall introduce an $\mathcal{H}$ quantity, this time referring to the velocity distribution f. It may be written $\mathcal{H} = \int f \log f \, dv$. Once again, this quantity can only decrease in time until equilibrium is reached and the velocity distribution becomes the equilibrium Maxwellian distribution.

In recent years there have been numerous numerical verifications of the uniform decrease of $\mathcal{H}$ with time. All of them confirm Boltzmann's prediction. Even today, his kinetic equation plays an important role in the physics of gases: transport coefficients such as those characterizing heat conductivity or diffusion can be calculated in good agreement with experimental data.

However, it is from the conceptual standpoint that Boltzmann's achievement is greatest: the distinction between reversible and irreversible phenomena, which, as we have seen, underlies the second law, is now transposed onto the microscopic level. The change of the velocity distribution due to free motion corresponds to the reversible part, while the contribution due to collisions corresponds to the irreversible part. For Boltzmann this was the key to the microscopic interpretation of entropy. A principle of molecular evolution had been produced! It is easy to understand the fascination this discovery exerted on the physicists who followed Boltzmann, including Planck, Einstein, and Schrödinger.[12]

Boltzmann's breakthrough was a decisive step in the direction of the physics of *processes.* What determines temporal evolution in Boltzmann's equation is no longer the Hamiltonian, depending on the type of forces; now, on the contrary, functions associated with the processes—for example, the cross section of scattering—will generate motion. Can we conclude that the problem of irreversibility has been solved, that Boltzmann's theory has reduced entropy to dynamics? The answer is clear: No, it has not. Let us have a closer look at this question.

Questioning Boltzmann's Interpretation

As soon as Boltzmann's fundamental paper appeared in 1872, objections were raised. Had Boltzmann really "deduced" irreversibility from dynamics? How could the reversible laws of trajectories lead to irreversible evolution? Is Boltzmann's kinetic equation in any way compatible with dynamics? It is easy to see that the symmetry present in Boltzmann's equation is in contradiction with the symmetry of classical mechanics.

We have already seen that velocity inversion ($v \rightarrow -v$) produces in classical dynamics the same effect as time inversion ($t \rightarrow -t$). This is a basic symmetry of classical dynamics, and we would expect that Boltzmann's kinetic equation, which describes the time change of the distribution function, would share this symmetry. But this is not so. The collision term calculated by Boltzmann remains *invariant with respect to velocity inversion.* There is a simple physical reason for this. Nothing in Boltzmann's picture distinguishes a collision that proceeds toward the future from a collision proceeding toward the past. This is the basis of Poincaré's objection to Boltzmann's derivation. A correct calculation can never lead to conclusions that contradict its premises.[13, 14] As we have seen, the symmetry properties of the kinetic equation obtained by Boltzmann for the distribution function contradict those of dynamics. Boltzmann cannot, therefore, have "deduced" entropy from dynamics. He must have introduced something new, something foreign to dynamics. Thus his results can represent at best only a phenomenological model that, however useful, has no direct relation with dynamics. This was also the objection that Zermelo (1896) brought against Boltzmann.

Loschmidt's objection, on the other hand, makes it possible to determine the *limits of validity* of Boltzmann's kinetic model. In fact, Loschmidt observed (1876) that this model can no longer be valid after a reversal of the velocities corresponding to the transformation $v \rightarrow -v$.

Let us explain this by means of a thought experiment. We start with a gas in a nonequilibrium condition and let it evolve till t_0. We then invert the velocities. The system reverts to its past state. As a consequence, Boltzmann's entropy is the same at $t=0$ and at $t=2t_0$.

We may multiply such thought experiments. Start with a mixture of hydrogen and oxygen; after some time water will appear. If we invert the velocities, we should go back to an initial state with hydrogen and oxygen and no water.

It is interesting that in laboratory or computer experiments, we actually can perform a velocity inversion. For example, in Figures 26 and 27, Boltzmann's $\mathcal{H}$ quantity has been calculated for two-dimensional hard spheres (hard disks), starting

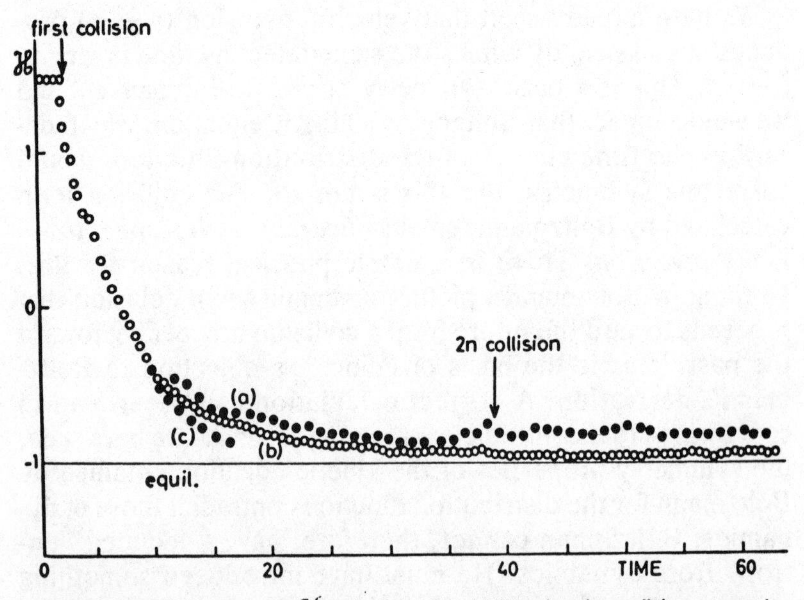

Figure 26. Evolution of $\mathcal{H}$ with time for N "hard spheres" by computer simulation; (a) corresponds to $N=100$, (b) to $N=484$, (c) to $N=1225$.

with disks on lattice sites with an isotropic velocity distribution. The results follow Boltzmann's predictions.

If, after fifty or a hundred collisions, corresponding to about 10^{-6} sec in a dilute gas, the velocities are inverted, a new ensemble is obtained.[15] Now, after the velocity inversion, Boltzmann's $\mathcal{H}$ quantity increases instead of decreasing.

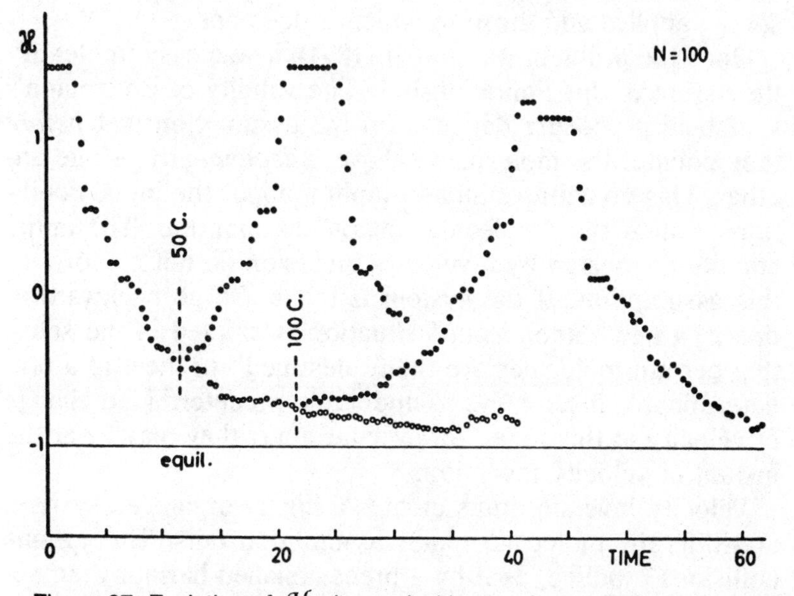

Figure 27. Evolution of $\mathcal{H}$ when velocities are inverted after 50 or 100 collisions. Simulation with 100 "hard spheres."

A similar situation can be produced in spin echo experiments or plasma echo experiments. There also, over limited periods of time, an "antithermodynamic" behavior in Boltzmann's sense may be observed.

But it is important to note that the velocity inversion experiment becomes increasingly more difficult when the time interval t_0 after which the inversion occurs is increased.

To be able to retrace its past, the gas must remember everything that happened to it during the time interval from 0 to t_0. There must be "storage" of information. We can express this storage in terms of correlations between particles. We shall come back to the question of correlations in Chapter IX. Let us only mention here that it is precisely this relation between correlations and collisions that is the element missing from

Boltzmann's considerations. When Loschmidt confronted him with this, Boltzmann had to accept that there was no way out: the collisions occurring in the opposite direction "undo" what was done previously, and the system has to revert to its initial state. Therefore, the function $\mathcal{H}$ must also increase until it again reaches its initial value. Velocity inversion thus calls for a distinction between the situations to which Boltzmann's reasoning applies and those to which it does not.

Once the problem was stated (1894), it was easy to identify the nature of this limitation.[16,17] The validity of Boltzmann's statistical procedure depends on the assumption that *before* they collide, the molecules behave *independently* of one another. This constitutes an assumption about the initial conditions, called the "molecular chaos" assumption. The initial conditions created by a velocity inversion do not conform to this assumption. If the system is made "to go backward in time," a new "anomalous" situation is created in the sense that certain molecules are then "destined" to meet at a predeterminable instant and to undergo a predetermined change of velocity at this time, however far apart they may be at the instant of velocity inversion.

Velocity inversion thus creates a highly organized system, and thus the molecular chaos assumption fails. The various collisions produce, as if by a preestablished harmony, an apparently purposeful behavior.

But there is more. What does the transition from order to disorder signify? In the Ehrenfest urn experiment, it is clear—the system will evolve till uniformity is reached. But other situations are not so clear; we may do computer experiments in which interacting particles are initially distributed at random. In time a lattice is formed. Do we still move from order to disorder? The answer is not obvious. To understand order and disorder we first have to define the objects in terms of which these concepts are used. Moving from dynamic to thermodynamic objects is easy in the case of dilute gases—as shown by the work of Boltzmann. However, it is not so easy in the case of dense systems whose molecules interact.

Because of such difficulties, Boltzmann's creative and pioneering work remained incomplete.

Dynamics and Thermodynamics: Two Separate Worlds

We already noted that trajectories are incompatible with the idea of irreversibility. However, the study of trajectories is not the only way in which we can give a formulation of dynamics. There is also the theory of ensembles introduced by Gibbs and Einstein,[6,18] which is of special interest in the case of systems formed by a large number of molecules. The essential new element in the Gibbs-Einstein ensemble theory is that we can formulate the dynamic theory independently of any precise specification of initial conditions.

The theory of ensembles represents dynamic systems in "phase space." The dynamic state of a point particle is specified by position (a vector with three components) and by momentum (also a vector with three components). We may represent this state by two points, each in a three-dimensional space, or by a single point in the six-dimensional space formed by the coordinates and momenta. This is the phase space. This geometric representation can be extended to an arbitrary system formed by n particles. We then need n x 6 numbers to specify the state of the system, or alternatively we may specify this system by a single point in the $6n$-dimensional phase space. The evolution in time of such a system will then be described by a trajectory in the phase space.

It has already been stated that the exact initial conditions of a macroscopic system are never known. Nevertheless, nothing prevents us from *representing* this system by an "ensemble" of points—namely, the points corresponding to the various dynamic states compatible with the information we have concerning the system. Each region of phase space may contain an infinite number of representative points, the density of which measures the probability of actually finding the system in this region. Instead of introducing an infinity of discrete points, it is more convenient to introduce a continuous density of representative points in the phase space. We shall call $\rho(q_1 \ldots q_{3n}, p_1 \ldots p_{3n})$ this density in phase space where $q_1, q_2 \ldots q_{3n}$ are the coordinates of the n points; similarly, $p_1, p_2 \ldots p_{3n}$ are the momenta (each point has three coordinates and three momenta). This density measures the probability of finding a dynamic system around the point $q_1 \ldots q_{3n}, p_1 \ldots p_{3n}$ in phase space.

Presented in such a way, the density function ρ may appear as an idealization, an artificial construct, whereas the trajectory of a point in phase space would correspond "directly" to the description of "natural" behavior. But in fact it is the point, not the density, that corresponds to an idealization. Indeed, we never know an initial state with the infinite degree of precision that would reduce a region in phase space to a single point; we can only determine an ensemble of trajectories starting from the ensemble of representative points corresponding to what we know about the initial state of the system. The density function ρ represents knowledge about a system, and the more accurate this knowledge, the smaller the region in the phase space where the density function is different from zero and where the system may be found. Should the density function everywhere have a uniform value, we would know nothing about the state of the system. It might be in any of the possible states compatible with its dynamic structure.

From this perspective, a point thus represents the *maximum knowledge* we can have about a system. It is the result of a limiting process, the result of the ever-growing precision of our knowledge. As we shall see in Chapter IX, a fundamental problem will be to determine when such a limiting process is really possible. Through increased precision, this process means we go from a region where the density function ρ is different from zero to another, smaller region inside the first. We can continue this until the region containing the system becomes arbitrarily small. But as we shall see, we must be cautious: arbitrarily small does not mean zero, and it is not certain a priori that this limiting process will lead to the possibility of consistently predicting a single well-defined trajectory.

The introduction of the theory of ensembles by Gibbs and Einstein was a natural continuation of Boltzmann's work. In this perspective the density function ρ in phase space replaces the velocity distribution function f used by Boltzmann. However, the physical content of ρ exceeds that of f. Just like f, the density function ρ determines the velocity distribution, but it also contains other information, such as the probability of meeting two particles a certain distance apart. In particular, correlations between particles, which we discussed in the preceding section, are now included in the density function ρ. In fact, this function contains the complete information about all statistical features of the *n*-body system.

We must now describe the evolution of the density function in phase space. At first sight, this appears to be an even more ambitious task than the one Boltzmann set himself for the velocity distribution function. But this is not the case. The Hamiltonian equations discussed in Chapter II allow us to obtain an exact evolution equation for ρ without any further approximations. This is the so-called Liouville equation, to which we shall return in Chapter IX. Here we wish merely to point out that the properties of Hamiltonian dynamics imply that the evolution of the density function ρ in phase space is that of an *incompressible fluid*. Once the representative points occupy a region of volume V in phase space, this volume remains *constant* in time. The shape of the region may be deformed in an arbitrary way, but the value of the volume remains the same.

Gibbs' theory of ensembles thus permits a rigorous combination of the statistical point of view (the study of the "population" described by ρ) and the laws of dynamics. It also permits a more accurate representation of the thermodynamic equilibrium state. Thus, in the case of an isolated system, the ensem-

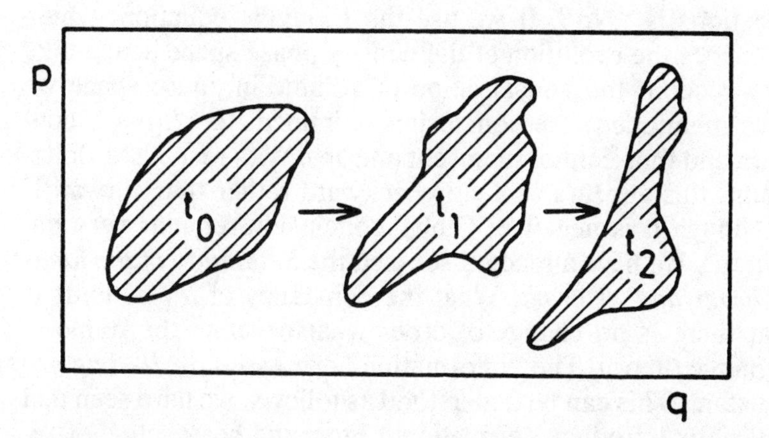

Figure 28. Time evolution in the phase space of a "volume" containing the representative points of a system: the volume is conserved while the shape is modified. The position in phase space is specified by coordinates *q* and momentum *p*.

ble of representative points corresponds to systems that all have the same energy E. The density ρ will differ from zero only on the "microcanonical surface" corresponding to the specified value of the energy in phase space. Initially, the density ρ may be distributed arbitrarily over this surface. At equilibrium, ρ must no longer vary with time and has to be independent of the specific initial state. Thus the approach to equilibrium has a simple meaning in terms of the evolution of ρ. The distribution function ρ becomes uniform over the microcanonical surface. Each of the points on this surface has the same probability of actually representing the system. This corresponds to the "microcanonical ensemble."

Does the theory of ensembles bring us any closer to the solution of the problem of irreversibility? Boltzmann's theory describes thermodynamic entropy in terms of the velocity distribution function f. He achieved this result through the introduction of his $\mathcal{H}$ quantity. As we have seen, the system evolves in time until the Maxwellian distribution is reached, while, during this evolution, the quantity $\mathcal{H}$ decreases uniformly. Can we now, in a more general fashion, take the evolution of the distribution ρ in phase space toward the microcanonical ensemble as the basis for entropy increase? Would it be enough to replace Boltzmann's quantity $\mathcal{H}$ expressed in terms of f by a "Gibbsian" quantity $\mathcal{H}_G$ defined in exactly the same way, but this time in terms of ρ? Unfortunately, the answer to both questions is "No." If we use the Liouville equation, which describes the evolution of the density phase space ρ, and take into account the conservation of volume in phase space we have mentioned, the conclusion is immediate: $\mathcal{H}_G$ is a constant and thus cannot represent entropy. With respect to Boltzmann, this appears as a step backward rather than forward!

Though it is negative, Gibbs' conclusion remains very important. We have already discussed the *ambiguity of the ideas of order and disorder.* What the constancy of $\mathcal{H}_G$ tells us is that there is no change of order whatsoever in the frame of dynamic theory! The "information" expressed by $\mathcal{H}_G$ remains constant. This can be understood as follows: we have seen that collisions introduce correlations. From the perspective of velocities, the result of collisions is randomization; therefore we can describe this process as a transition from order to disorder, but the appearance of correlations as the result of collision points in the opposite direction, toward a transition from disorder to order! Gibbs' result shows that the two effects exactly cancel each other.

We come, therefore, to an important conclusion. Whatever representation we use, be it the idea of trajectories or the Gibbs-Einstein ensemble theory, we will never be able to deduce a theory of irreversible processes that will be valid for every system that satisfies the laws of classical (or quantum) dynamics. There isn't even a way to speak of a transition from order to disorder! How should we understand these negative results? Is any theory of irreversible processes in absolute conflict with dynamics (classical or quantum)? It has often been proposed that we include some cosmological terms that would express the influence of the expanding universe on the equations of motion. Cosmological terms would ultimately provide the arrow of time. However, this is difficult to accept. On the one hand, it is not clear how we should add these cosmological terms; on the other, precise dynamic experiments seem to rule out the existence of such terms, at least on the terrestrial scale with which we are concerned here (think, for example, about the precision of space trip experiments, which confirm Newton's equations to a high degree). On the other hand, as we have already stated, we live in a pluralistic universe in which reversible and irreversible processes coexist, all embedded in the expanding universe.

An even more radical conclusion is to affirm with Einstein that time as irreversibility is an illusion that will never find a place in the objective world of physics. Fortunately there is another way out, which we shall explore in Chapter IX. Irreversibility, as has been repeatedly stated, is not a universal property. Therefore, no general derivation of irreversibility from dynamics is to be expected.

Gibbs' theory of ensembles introduces only one additional element with respect to trajectory dynamics, but a very important one—our ignorance of the precise initial conditions. It is unlikely that this ignorance alone leads to irreversibility.

We should therefore not be astonished at our failure. We have not yet formulated the specific features that a dynamic system has to possess to lead to irreversible processes.

Why have so many scientists accepted so readily the subjective interpretation of irreversibility? Perhaps part of its attraction lies in the fact that, as we have seen, the irreversible increase of entropy was at first associated with imperfect manipulation, with our lack of control over operations that are ideally reversible.

But this interpretation becomes absurd as soon as the irrelevant associations with technological problems are set aside. We must remember the context that gave the second law its significance as nature's arrow of time. According to the subjective interpretation, chemical affinity, heat conduction, viscosity, all the properties connected with irreversible entropy production would depend on the observer. Moreover, the extent to which phenomena of organization originating in irreversibility play a role in biology makes it impossible to consider them as simple illusions due to our ignorance. Are we ourselves—living creatures capable of observing and manipulating—mere fictions produced by our imperfect senses? Is the distinction between life and death an illusion?

Thus recent developments in thermodynamic theory have increased the violence of the conflict between dynamics and thermodynamics. Attempts to reduce the results of thermodynamics to mere approximations due to our imperfect knowledge seem wrongheaded when the constructive role of entropy is understood and the possibility of an amplification of fluctuations is discovered. Conversely, it is difficult to reject dynamics in the name of irreversibility: there is no irreversibility in the motion of an ideal pendulum. Apparently there are two conflicting worlds, a world of trajectories and a world of processes, and there is no way of denying one by asserting the other.

To a certain extent, there is an analogy between this conflict and the one that gave rise to dialectical materialism. We have described in Chapters V and VI a nature that might be called "historical"—that is, capable of development and innovation. The idea of a history of nature as an integral part of materialism was asserted by Marx and, in greater detail, by Engels. Contemporary developments in physics, the discovery of the constructive role played by irreversibility, have thus raised within the natural sciences a question that has long been asked by materialists. For them, understanding nature meant understanding it as being capable of producing man and his societies.

Moreover, at the time Engels wrote his *Dialectics of Nature*, the physical sciences seemed to have rejected the mechanistic world view and drawn closer to the idea of an historical development of nature. Engels mentions three fundamental discoveries: energy and the laws governing its qualitative transformations, the cell as the basic constituent of life, and Darwin's discovery of the evolution of species. In view of these great discoveries, Engels came to the conclusion that the mechanistic world view was dead.

But mechanicism remained a basic difficulty facing dialectical materialism. What are the relations between the general laws of dialectics and the equally universal laws of mechanical motion? Do the latter "cease" to apply after a certain stage has been reached, or are they simply false or incomplete? To come back to our previous question, how can the world of processes and the world of trajectories ever be linked together?[19]

However, while it is easy to criticize the subjectivistic interpretation of irreversibility and to point out its weakness, it is not so easy to go beyond it and formulate an "objective" theory of irreversible processes. The history of this subject has some dramatic overtones. Many people believe that it is the recognition of the basic difficulties involved that may have led to Boltzmann's suicide in 1906.

Boltzmann and the Arrow of Time

As we have noted, Boltzmann at first thought that he could prove that the arrow of time was determined by the evolution of dynamic systems toward states of higher probability or a higher number of complexions: there would be a one-way increase of the number of complexions with time. We have also discussed the objections of Poincaré and Zermelo. Poincaré proved that every closed dynamic system reverts in time toward its previous state. Thus, all states are forever recurring. How could such a thing as an "arrow of time" be associated with entropy increase? This led to a dramatic change in Boltzmann's attitude. He abandoned his attempt to prove that an objective arrow of time exists and introduced instead an idea that, in a sense, reduced the law of entropy increase to a tautology. Now he argued that the arrow of time is only a convention that we (or perhaps all living beings) introduce into a world in which there is no objective distinction between past and future. Let us cite Boltzmann's reply to Zermelo:

> We have the choice of two kinds of picture. Either we assume that the whole universe is at the present moment in a very improbable state. Or else we assume that the aeons during which this improbable state lasts, and the distance from here to Sirius, are minute if compared with the age and size of the whole universe. In such a universe, which is in thermal equilibrium as a whole and therefore dead, relatively small regions of the size of our galaxy will be found here and there; regions (which we may call "worlds") which deviate significantly from thermal equilibrium for relatively short stretches of those "aeons" of time. Among these worlds the probabilities of their state (i.e. the entropy) will increase as often as they decrease. In the universe as a whole the two directions of time are indistinguishable, just as in space there is no up or down. However, just as at a certain place on the earth's surface we can call "down" the direction towards the centre of the earth, so a living organism that finds itself in such a world at a certain period of time can define the "direction" of time as going from the less probable state to the more probable one (the former will be the "past" and the latter the "future"), and by virtue of this definition he will find that his own small region, isolated from the rest of the universe, is "initially" always in an improbable state. It seems to me that this way of looking at things is the only one which allows us to understand the validity of the second law, and the heat death of each individual world, without invoking a unidirectional change of the entire universe from a definite initial state to a final state.[20]

Boltzmann's idea can be made clearer by referring to a diagram proposed by Karl Popper (Figure 29). The arrow of time would be as arbitrary as the vertical direction determined by the gravitational field.

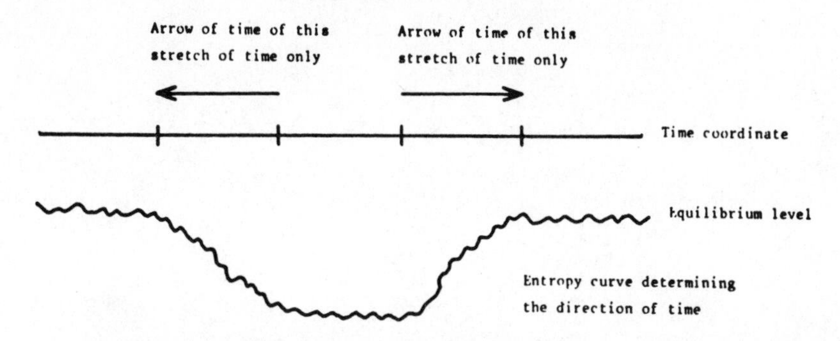

Figure 29. Popper's schematic representation of Boltzmann's cosmological interpretation of the arrow of time (see text).

Commenting on Boltzmann's text, Popper has written:

> I think that Boltzmann's idea is staggering in its boldness and beauty. But I also think that it is quite untenable, at least for a realist. It brands unidirectional change as an illusion. This makes the catastrophe of Hiroshima an illusion. Thus it makes our world an illusion, and with it all our attempts to find out more about our world. It is therefore self-defeating (like every idealism). Boltzmann's idealistic ad hoc hypothesis clashes with his own realistic and almost passionately maintained anti-idealistic philosophy, and with his passionate wish to know.[21]

We fully agree with Popper's comments, and we believe that it is time to take up Boltzmann's task once again. As we have said, the twentieth century has seen a great conceptual revolution in theoretical physics, and this has produced new hopes for the unification of dynamics and thermodynamics. We are now entering a new era in the history of time, an era in which both being and becoming can be incorporated into a single noncontradictory vision.

CHAPTER IX

IRREVERSIBILITY—THE ENTROPY BARRIER

Entropy and the Arrow of Time

In the preceding chapter we described some difficulties in the microscopic theory of irreversible processes. Its relation with dynamics, either classical or quantum, cannot be simple, in the sense that irreversibility and its concomitant increase of entropy cannot be a general consequence of dynamics. A microscopic theory of irreversible processes will require additional, more specific conditions. We must accept a pluralistic world in which reversible and irreversible processes coexist. Yet such a pluralistic world is not easy to accept.

In his *Dictionnaire Philosophique* Voltaire wrote the following about destiny:

> . . . everything is governed by immutable laws . . . everything is prearranged . . . everything is a necessary effect. . . . There are some people who, frightened by this truth, allow half of it, like debtors who offer their creditors half their debt, asking for more time to pay the remainder. There are, they say, events which are necessary and others which are not. It would be strange if a part of what happens had to happen and another part did not. . . . I necessarily must have the passion to write this, and you must have the passion to condemn me; we are both equally foolish, both toys in the hands of destiny. Your nature is to do ill, mine is to love truth, and to publish it in spite of you.[1]

However convincing they may sound, such a priori arguments can lead us astray. Voltaire's reasoning is Newtonian: nature always conforms to itself. But, curiously, today we find ourselves in the strange world mocked by Voltaire; we are astonished to discover the qualitative diversity presented by nature.

It is not surprising that people have vacillated between the two extremes; between the elimination of irreversibility from physics, advocated by Einstein, as we have mentioned,[2] and, on the contrary, the emphasis on the importance of irreversibility, as in Whitehead's concept of process. There can be no doubt that irreversibility exists on the macroscopic level and has an important constructive role, as we have shown in Chapters V and VI. Therefore there must be something in the microscopic world of which macroscopic irreversibility is the manifestation.

The microscopic theory has to account for two closely related elements. First of all, we must follow Boltzmann in attempting to construct a microscopic model for entropy (Boltzmann's $\mathcal{H}$ function) that changes uniformly in time. This change has to define our arrow of time. The increase of entropy for isolated systems has to express the aging of the system.

Often we may have an arrow of time without being able to associate entropy with the type of processes considered. Popper gives a simple example of a system presenting a unidirectional process and therefore an arrow of time.

> Suppose a film is taken of a large surface of water initially at rest into which a stone is dropped. The reversed film will show contracting, circular waves of increasing amplitude. Moreover, immediately behind the highest wave crest, a circular region of undisturbed water will close in towards the centre. This cannot be regarded as a possible classical process. It would demand a vast number of distant coherent generators of waves the coordination of which, to be explicable, would have to be shown, in the film, as originating from one centre. This, however, raises precisely the same difficulty again, if we try to reverse the amended film.[3]

Indeed, whatever the technical means, there will always be a distance from the center beyond which we are unable to generate a contracting wave. There are unidirectional processes. Many other processes of the type presented by Popper can be imagined: we never see energy coming from all directions converge on a star, together with the backward-running nuclear reactions that would absorb that energy.

In addition, there exist other arrows of time—for example, the cosmological arrow (see the excellent account by M. Gardner[4]). If we assume that the universe started with a Big Bang, this obviously implies a temporal order on the cosmological level. The size of the universe continues to increase, but we cannot identify the radius of the universe with entropy. Indeed, as we already mentioned, inside the expanding universe we find both reversible and irreversible processes. Similarly, in elementary-particle physics there exist processes that present the so-called *T*-violation. The *T*-violation implies that the equations describing the evolution of the system for $+t$ are different from those describing the evolution for $-t$. However, this *T*-violation does not prevent us from including it in the usual (Hamiltonian) formulation of dynamics. No entropy function can be defined as a result of the *T*-violation.

We are reminded of the celebrated discussion between Einstein and Ritz published in 1909.[5] This is a quite unusual paper, a very short one, less than a printed page long. It simply is a statement of disagreement. Einstein argued that irreversibility is a consequence of the probability concept introduced by Boltzmann. On the contrary, Ritz argued that the distinction between "retarded" and "advanced" waves plays an essential role. This distinction reminds us of Popper's argument. The waves we observe in the pond are retarded waves; they follow the dropping of the stone.

Both Einstein and Ritz introduced essential elements into the discussion of irreversibility, but each of them emphasized only part of the story. We have already mentioned in Chapter VIII that probability *presupposes* a direction of time and therefore cannot be used to derive the arrow of time. We have also mentioned that the exclusion of processes such as advanced waves does not necessarily lead to a formulation of the second law. We need both types of arguments.

Irreversibility as a Symmetry-Breaking Process

Before discussing the problem of irreversibility, it is useful to remember how another type of symmetry-breaking, spatial symmetry-breaking, can be derived. In the equations describing reaction-diffusion systems, left and right play the same role (the diffusion equations remain invariant when we perform the space inversion $r \rightarrow -r$). Still, as we have seen, bifurcations may lead to solutions in which this symmetry is broken (see Chapter V). For example, the concentration of some of the components may become higher on the left than on the right. The symmetry of the equations only requires that the symmetry-breaking solutions appear in pairs.

There are, of course, many reaction-diffusion equations that do not present bifurcations and that therefore cannot break spatial symmetry. The breaking of spatial symmetry requires other highly specific conditions. This is valuable for understanding temporal symmetry-breaking, in which we are primarily interested here. We have to find systems in which the equations of motion may have realizations of *lower symmetry.*

The equations are indeed invariant in respect to time inversion $t \rightarrow -t$. However, the realization of these equations may correspond to evolutions that lose this symmetry. The only condition imposed by the symmetry of equations is that such realizations appear in pairs. If, for example, we find one solution going to equilibrium in the far distant future (and not in the far distant past), we should also find a solution that goes to equilibrium in the far distant past (and not in the far distant future). Symmetry-broken solutions appear in pairs.

Once we find such a situation we can express the intrinsic meaning of the second law. It becomes a selection principle stating that only one of the two types of solutions can be realized or may be observed in nature. Whenever applicable, the second law expresses an intrinsic polarization of nature. It can never be the outcome of dynamics itself. It has to appear as a supplementary selection principle that when realized is propagated by dynamics. Only a few years ago it seemed impossible to attempt such a program. However, over the past few decades dynamics has made remarkable progress, and we can now understand in detail how these symmetry-breaking solutions emerge in dynamic systems "of sufficient complexity" and what the selection rule expressed by the second law of thermodynamics means on the microscopic level. This is what we want to show in the next part of this chapter.

The Limits of Classical Concepts

Let us start with classical mechanics. As we have already mentioned, if trajectory is to be the basic irreducible element, the world would be as reversible as the trajectories out of which it is formed. In this description there would be no entropy and no arrow of time; but, as a result of unexpected recent developments, the validity of the trajectory concept appears far more limited than we might have expected. Let us return to Gibbs' and Einstein's theory of ensembles, introduced in Chapter VIII. We have seen that Gibbs and Einstein introduced phase space into physics to account for the fact that we do not "know" the initial state of systems formed by a large number of particles. For them, the distribution function in phase space was only an auxiliary construction expressing our *de facto* ignorance of a situation that was well determined *de jure*. However, the entire problem takes on new dimensions once it can be shown that *for certain types of systems* an infinitely precise determination of initial conditions leads to a self-contradictory procedure. Once this is so, the fact that we never know a single trajectory but a group, an ensemble of trajectories in phase space, is not a mere expression of the limits of our knowledge. It becomes a starting point of a new way of investigating dynamics.

It is true that in simple cases there is no problem. Let us take the example of a pendulum. It may oscillate or else rotate about its axis according to the initial conditions. For it to rotate, its kinetic energy must be large enough for it not to "fall back" before reaching a vertical position. These two types of motion define two disjointed regions of phase space. The reason for this is very simple: rotation requires more energy than oscillation (see Figure 30).

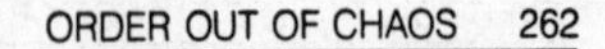

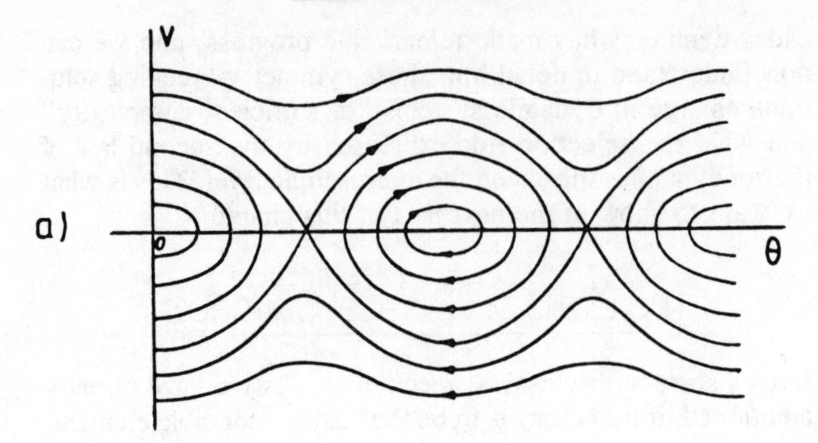

V
b)
0°
360°
θ

Figure 30. Representation of a pendulum's motion in a space where V is the velocity and Θ the angle of deflection. (a) typical trajectories in (V, Θ) space; (b) the shaded regions correspond to oscillations; the region outside corresponds to rotations.

If our measurements allow us to establish that the system is initially in a given region, we may safely predict the type of motion displayed by the pendulum. We can increase the accuracy of our measurements and localize the initial state of the pendulum in a smaller region circumscribed by the first. In any case, we know the system's behavior for all time; nothing new or unexpected is likely to occur.

One of the most surprising results achieved in the twentieth century is that *such a description is not valid in general.* On the contrary, "most" *dynamic systems* behave in a quite unstable way.[6] Let us indicate one kind of trajectory (for example, that of oscillation) by + and another kind (for example, that corresponding to rotation) by *. Instead of Figure 30, where the two regions were separated, we find, in general, a mixture of states that makes the transition to a single point ambiguous (see Figure 31). If we know only that the initial state of our system is in region A, we cannot deduce that its trajectory is of type +; it may equally well be of type *. We achieve nothing by increasing the accuracy by going from region A to a smaller region within it, for the uncertainty remains. *In all regions, however small, there are always states belonging to each of the two types of trajectories.*[7]

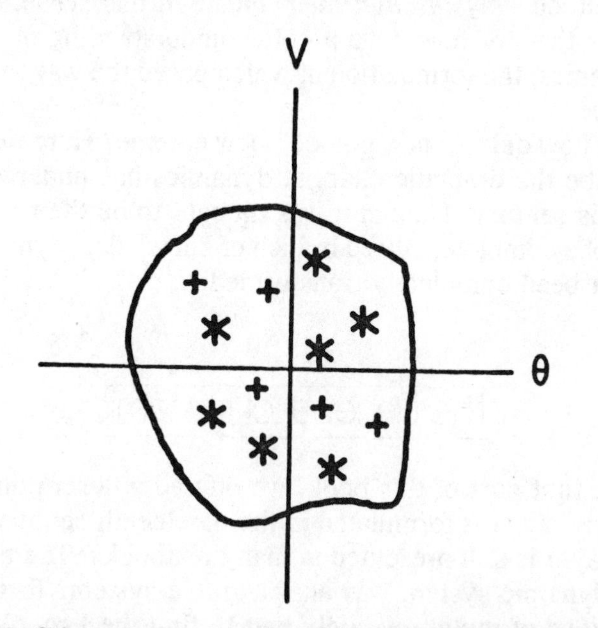

Figure 31. Schematic representation of any region, arbitrarily small, of the phase space A for a system presenting dynamic instability. As in the case of the pendulum, there are two types of trajectories (represented here by + and *); however, in contrast with the pendulum, both motions appear in every region arbitrary small.

For such systems, a trajectory becomes *unobservable*. This instability expresses the limits of Newtonian idealization. The independence of the two basic elements of Newtonian dynamics, the dynamic law and the initial conditions, is destroyed: the dynamic law enters into conflict with the determination of the initial conditions. We may recall the way in which Anaxagoras conceived of the wealth of nature's creative possibilities. For him, every object contains in each of its parts an infinite multiplicity of qualitatively different seeds. Here also, each region of phase space maintains a wealth of qualitatively different behaviors.

In this perspective, the deterministic trajectory appears to have limited application. Inasmuch as we are incapable, not only in practice but also in theory, of describing a system by means of a trajectory and are *obliged* to use a distribution function corresponding to a finite region (however small) of phase space, we can only predict the *statistical* future of the system.

Our friend Leon Rosenfeld used to say that concepts can be understood only *through their limits*. In this sense, it would appear that we now have a better understanding of classical mechanics, the formulation of which paved the way to modern science.

But how did this new point of view emerge? Here we have to describe the dramatic changes dynamics has undergone during this century. Though it was thought to be the very archetype of a complete, closed branch of knowledge, dynamics has in fact been completely transformed.

The Renewal of Dynamics

In the first part of this book, we offered a description of dynamics as it was formulated in the nineteenth century. This is the way it is still presented in many textbooks. The prototype of a dynamic system was an integrable system. To solve the equations of motion we only had to find the "good" coordinates, such that the corresponding moments were invariants of motion. In this way interactions between moving entities were eliminated. This program failed. We have already mentioned that at the end of the nineteenth century Bruns and Poincaré demonstrated that most dynamic systems, starting with the famous "three body" problem, were not integrable.

On the other hand, the very idea of approaching equilibrium in terms of the theory of ensembles requires that we go beyond the idealization of integrable systems. As we saw in Chapter VIII, according to the theory of ensembles, an isolated system is in equilibrium when it is represented by a "microcanonical ensemble," when all points on the surface of given energy have the same probability. This means that for a system to evolve to equilibrium, energy must be the only quantity conserved during its evolution. It must be the only "invariant." Whatever the initial conditions, the evolution of the system must allow it to reach all points on the surface of given energy. But for an integrable system, energy is far from being the only invariant. In fact, there are as many invariants as degrees of freedom, since each generalized momentum remains constant. Therefore we have to expect that such a system is "imprisoned" in a very small "fraction" of the constant-energy (see Figure 32) surface formed by the intersection of all these invariant surfaces.

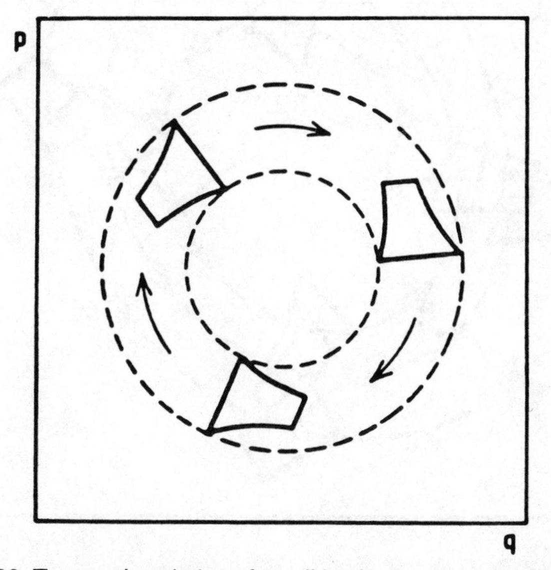

Figure 32. Temporal evolution of a cell in phase space p, q. The "volume" of the cell and its form are maintained in time; moreover, most of the phase space is inaccessible to the system.

To avoid these difficulties, Maxwell and Boltzmann introduced a new, quite different type of dynamic system. For these systems energy would be the only invariant. Such systems are called "ergodic" systems (see Figure 33).

Great contributions to the theory of ergodic systems have been made by Birkhoff, von Neumann, Hopf, Kolmogoroff, and Sinai, to mention only a few.[8,9,10] Today we know that there are large classes of dynamic (though non-Hamiltonian) systems that are ergodic. We also know that even relatively simple systems may have properties stronger than ergodicity. For these systems, motion in phase space becomes highly chaotic (while always preserving a volume that agrees with the Liouville equation mentioned in Chapter VII).

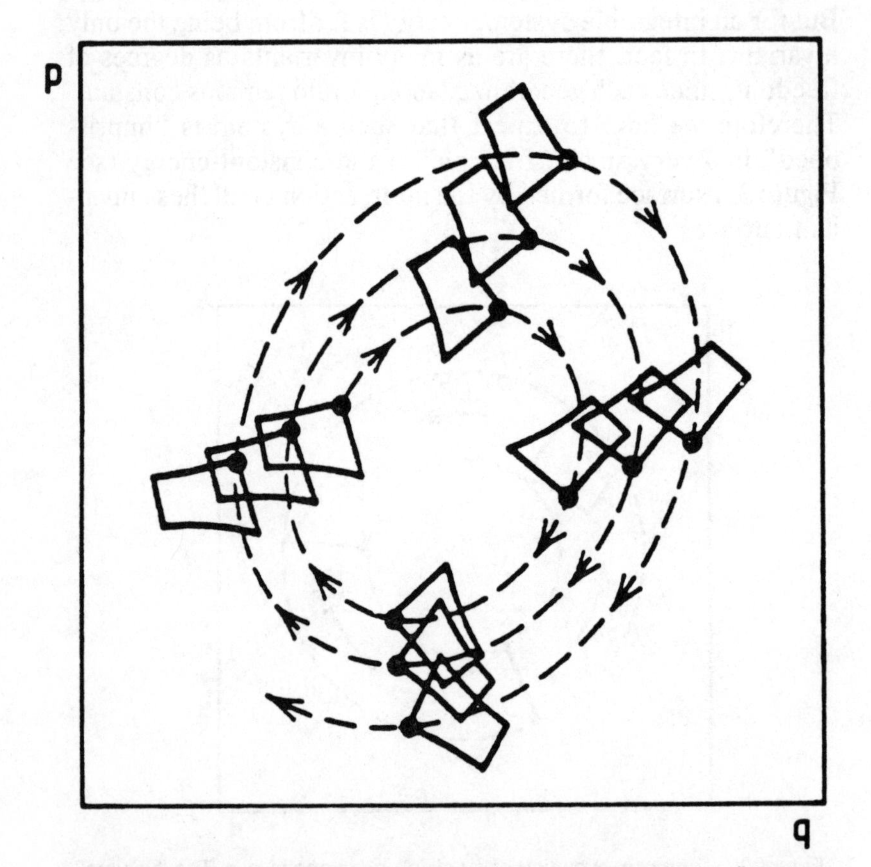

Figure 33. Typical evolution in phase space of a cell corresponding to an ergodic system. Time going on, the "volume" and the form are conserved but the cell now spirals through the whole phase space.

Suppose that our knowledge of initial conditions permits us to localize a system in a small cell of the phase space. During its evolution, we shall see this initial cell twist and turn and, like an amoeba, send out "pseudopods" in all directions, spreading out in increasingly thinner and ever more twisted filaments until it finally invades the whole space. No sketch can do justice to the complexity of the actual situation. Indeed, during the dynamic evolution of a mixing system, two points as close together in phase space as we might wish may head in different directions. Even if we possess a lot of information about the system so that the initial cell formed by its representative points is very small, dynamic evolution turns this cell into a true geometric "monster" stretching its network of filaments through phase space.

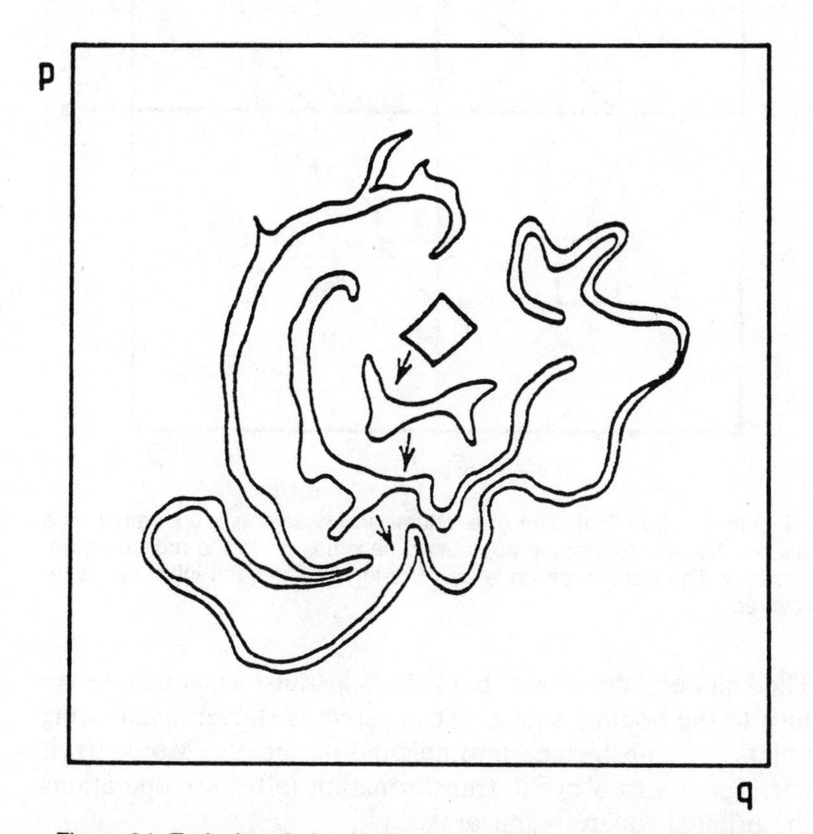

Figure 34. Typical evolution in phase space of a cell corresponding to a "mixing" system. The volume is still conserved but no more its form: the cell progressively spreads through the whole phase space.

We would like to illustrate the distinction between stable and unstable systems with a few simple examples. Consider a phase space with two dimensions. At regular time intervals, we shall replace these coordinates by new ones. The new point on the horizontal axis is p-q, the new ordinate p. Figure 35 shows what happens when we apply this operation to a square.

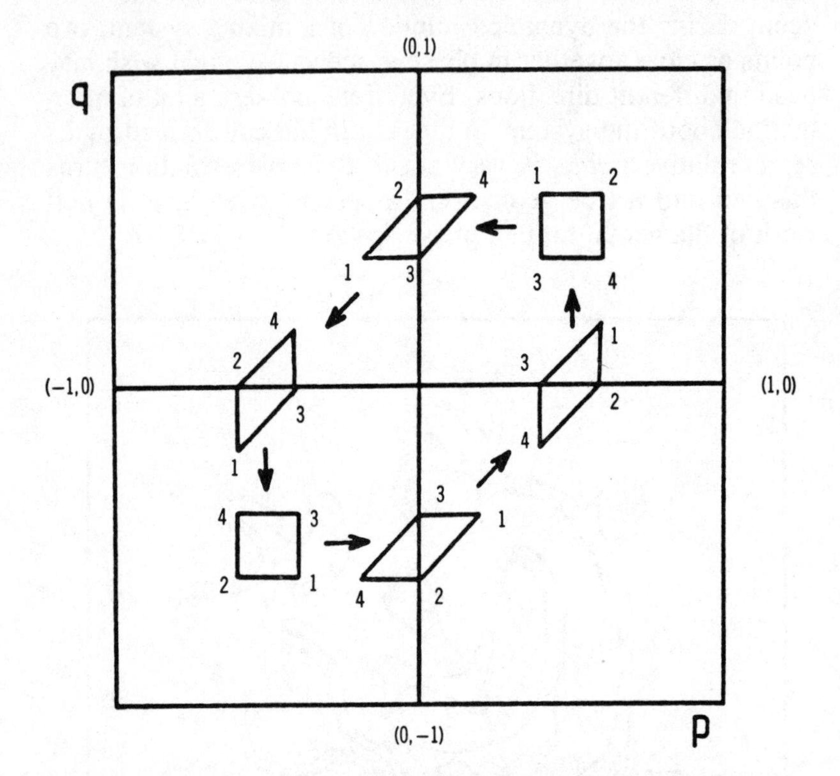

Figure 35. Transformation of a volume in phase space generated by a discrete transformation: the abscissa p becomes $p-q$, the ordinate q becomes p. The transformation is cyclic: after six times the initial cell is recovered.

The square is deformed, but after six transformations we return to the original square. The system is stable: neighboring points are transformed into neighboring points. Moreover, it corresponds to a cyclic transformation (after six operations the original square reappears).

Let us now consider two examples of highly unstable systems—the first mathematical, the second of obvious physical relevance. The first system consists of a transformation that, for obvious reasons, mathematicians call the "baker transformation."[9, 10] We take a square and flatten it into a rectangle, then we fold half of the rectangle over the other half to form a square again. This set of operations is shown in Figure 36 and may be repeated as many times as one likes.

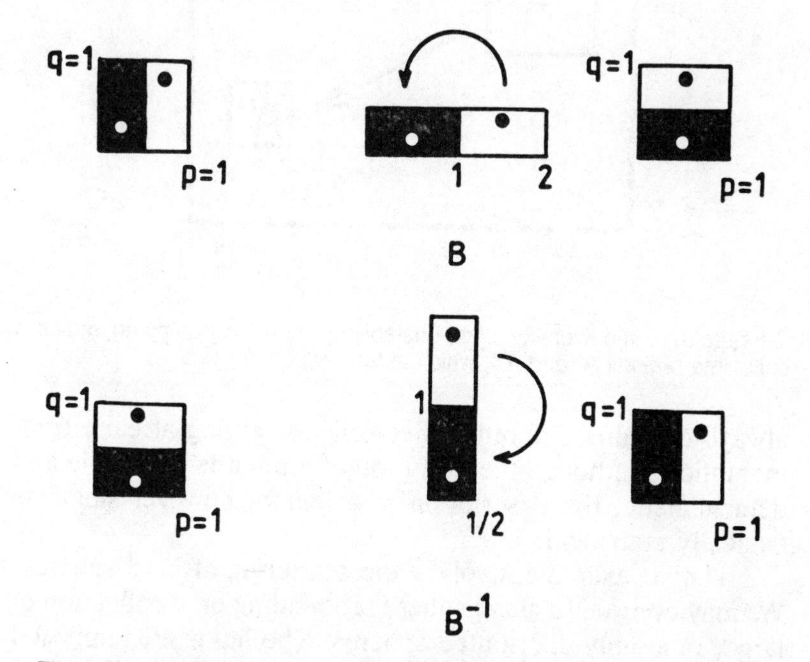

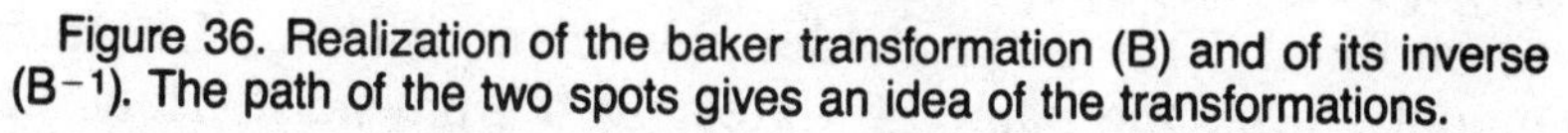

Figure 36. Realization of the baker transformation (B) and of its inverse (B^{-1}). The path of the two spots gives an idea of the transformations.

Each time the surface of the square is broken up and redistributed. The square corresponds here to the phase space. The baker transformation transforms each point into a well-defined new point. Although the series of points obtained in this way is "deterministic," the system displays in addition irreducibly statistical aspects. Let us take, for instance, a system described by an initial condition such that a region A of the square is initially filled in a uniform way with representative points. It may be shown that after a sufficient number of repetitions of the transformation, this cell, *whatever its size and localization,* will be broken up into pieces (see Figure 37). The essential point is that any region, whatever its size, thus

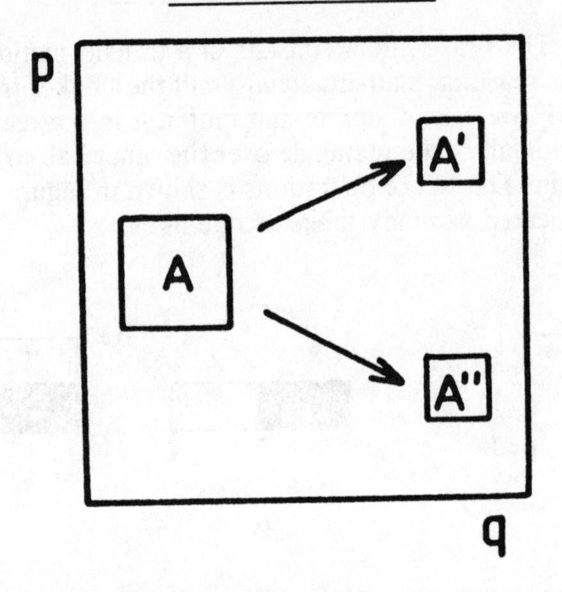

Figure 37. Time evolution of an unstable system. Time going on, region A splits into regions A′ and A″, which in turn will be divided.

always contains different trajectories diverging at each fragmentation. Although the evolution of a point is reversible and deterministic, the description of a region, however small, is basically statistical.

A similar example involves the scattering of hard spheres. We may consider a small sphere rebounding on a collection of large, randomly distributed spheres. The latter are supposed to be fixed. This is the model physicists call the "Lorentz model," after the name of a great Dutch physicist, Hendrik Antoon Lorentz.

The trajectory of the small mobile sphere is well defined. However, whenever we introduce the smallest uncertainty in the initial conditions, this uncertainty is amplified through successive collisions. As time passes, the probability of finding the small sphere in a given volume becomes uniform. Whatever the number of transformations, we never return to the original state.

In the last two examples we have strongly *unstable* dynamic systems. The situation is reminiscent of instabilities as they appear in thermodynamic systems (see Chapter V). Arbitrarily small differences in initial conditions are amplified. As a result we can no longer perform the transition from ensembles

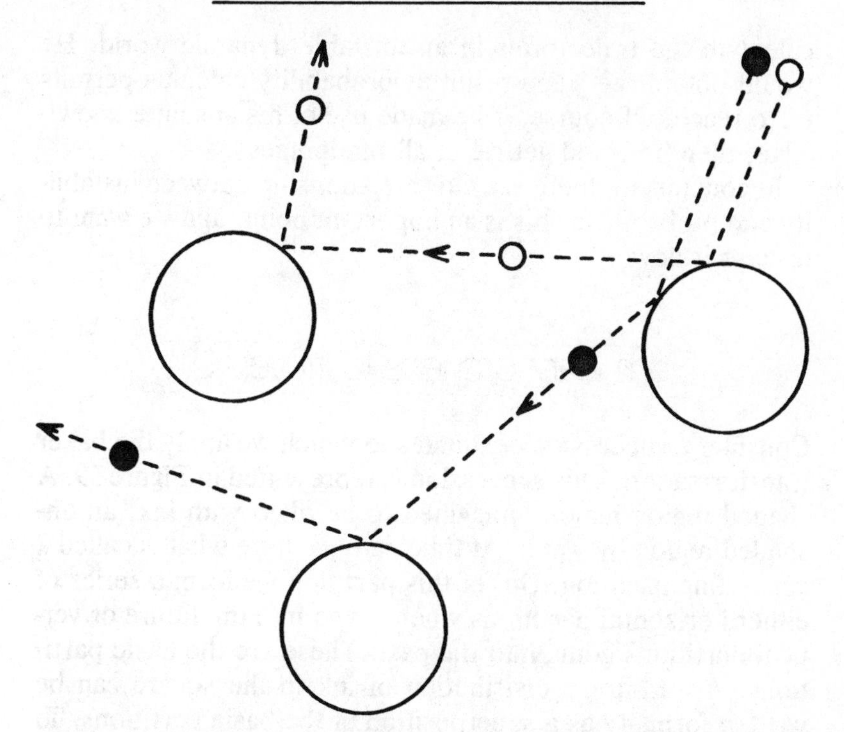

Figure 38. Schematic representation of the instability of the trajectory of a small sphere rebounding on large ones. The least imprecision about the position of the small sphere makes it impossible to predict which large sphere it will hit after the first collision.

in phase space to individual trajectories. The description in terms of ensembles has to be taken as the starting point. Statistical concepts are no longer merely an approximation with respect to some "objective truth." When faced with these unstable systems, Laplace's demon is just as powerless as we.

Einstein's saying, "God does not play dice," is well known. In the same spirit Poincaré stated that for a supreme mathematician there is no place for probabilities. However, Poincaré himself mapped the path leading to the answer to this problem.[11] He noticed that when we throw dice and use probability calculus, it does not mean that we suppose dynamics to be wrong. It means something quite different. We apply the probability concept because in each interval of initial conditions, however small, there are as "many" trajectories that lead to each of the faces of the dice. This is precisely what happens with unstable dynamic systems. God could, if he wished to, calculate the trajectories in an unstable dynamic world. He would obtain the same result as probability calculus permits us to reach. Of course, if he made use of his absolute knowledge, then he could get rid of all randomness.

In conclusion, there is a close relationship between instability and probability. This is an important point, and we want to discuss it now.

From Randomness to Irreversibility

Consider a succession of squares to which we apply the baker transformation. This succession is represented in Figure 39. A shaded region may be imagined to be filled with ink, an unshaded region by water. At time zero we have what is called a generating partition. Out of this partition we form a series of either horizontal partitions when we go into the future or vertical partitions going into the past. These are the basic partitions. An arbitrary distribution of ink in the square can be written formally as a superposition of the basic partitions. To each basic partition we may associate an "internal" time that is simply the number of baker transformations we have to perform to go from the generating partition to the one under consideration.[12] We therefore see that this type of system admits a kind of internal age.*

The internal time *T* is quite different from the usual mechanical time, since it depends on the global topology of the system. We may even speak of the "timing of space," thus coming close to ideas recently put forward by geographers, who have introduced the concept of "chronogeography."[13] When we look at the structure of a town, or of a landscape, we see temporal elements interacting and coexisting. Brasília or Pompeii would correspond to a well-defined internal age, somewhat like one of the basic partitions in the baker transformation. On the contrary, modern Rome, whose buildings originated in quite dif-

*It may be noticed that this internal time, which we shall denote by *T*, is in fact an operator like those introduced in quantum mechanics (see Chapter VII). Indeed, an arbitrary partition of the square does not have a well-defined time but only an "average" time corresponding to the superposition of the basic partitions out of which it is formed.

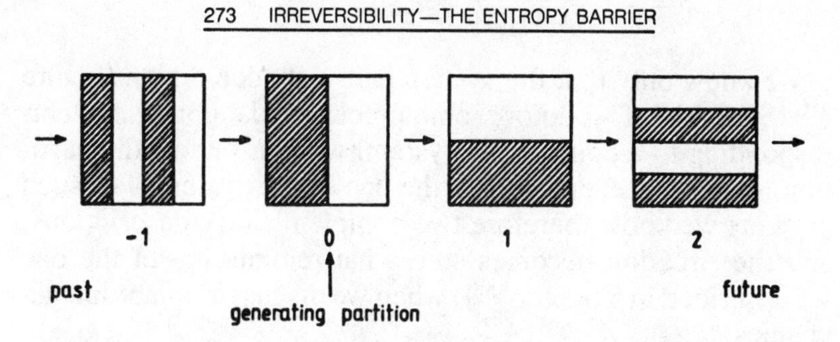

Figure 39. Starting with the "generating partition" (see text) at time 0, we repeatedly apply the baker transformation. We generate horizontal stripes in this way. Similarly going into the past we obtain vertical stripes.

ferent periods, would correspond to an average time exactly as an arbitrary partition may be decomposed into elements corresponding to different internal times.

Let us again look at Figure 39. What happens if we go into the far distant future? The horizontal bands of ink will get closer and closer. Whatever the precision of our measurements, after some time we shall conclude that the ink is uniformly distributed over the volume. It is therefore not surprising that this kind of approach to "equilibrium" may be mapped into a stochastic process, such as the Markov chain we described in Chapter VIII. Recently this has been shown with full mathematical rigor,[14] but the result seems to us quite natural. As time passes, the distribution of ink reaches equilibrium, exactly like the distribution of balls in the urn experiment discussed in Chapter VIII. However, when we look into the past, again beginning from the generating partition at time zero, we see the same phenomenon. Now ink is distributed along shrinking vertical sections and, again, sufficiently far in the past we shall find a uniform distribution of ink. We may therefore conclude that we can also model this process in terms of a Markov chain, now, however, oriented toward the past. We see that out of the unstable dynamic processes we obtain two Markov chains, one reaching equilibrium in the future, one in the past.

We believe that it is a very interesting result, and we would like to comment it. Internal time provides us with a new 'nonlocal' description.

When we know the 'age' of the system, (that is, the corresponding partition), we can still not associate to it a well-defined local trajectory.

We know only that the system is in a shaded region (Figure 39). Similarly, if we know some precise initial conditions corresponding to a point in the system, we don't know the partition to which it belongs, nor the age of the system. For such systems we know therefore two complementary descriptions, and the situation becomes somewhat reminiscent of the one we described in Chapter VII, when we discussed quantum mechanics.

It is because of the existence of this new alternative, non-local description, that we can make the transition from dynamics to probabilities. We call the systems for which this is possible "intrinsically random systems".

In classical deterministic systems, we may use transition probabilities to go from one point to another on a quite degenerate sense. This transition probability will be equal to one if the two points lie on the same dynamic trajectory, or zero if they are not.

In contrast, in genuine probability theory, we need transition probabilities which are positive numbers between zero and one. How is this possible? Here we see in full light the conflict between subjectivistic views and objective interpretations of probability. The subjective interpretation corresponds to the situation where individual trajectories are not known. Probability (and, eventually, irreversibility, closely related to it) would originate from our ignorance. But fortunately, there is another objective interpretation: probability arises as a result of an alternative description of dynamics, a non-local description which arises in strongly unstable dynamical systems.

Here, probability becomes an objective property generated from the inside of dynamics, so to speak, and which expresses a basic structure of the dynamical system. We have stressed the importance of Boltzmann's basic discovery: the connection between entropy and probability. For intrinsic random systems, the concept of probability acquires a dynamical meaning. We have now to make the transition from intrinsic random systems to irreversible systems. We have seen that out of unstable dynamical processes, we obtain two Markov chains.

We may see this duality in a different way. Take a distribution concentrated on a line (instead of being distributed on a surface). This line may be vertical or horizontal. Let us look at what will happen to this line when we apply the baker transformation going to the future. The result is represented in Figure 40. The vertical line is cut successively into pieces and will reduce to a point in the far distant future. The horizontal line, on the contrary, is duplicated and will uniformly "cover" the surface in the far distant future. Obviously, just the opposite happens if we go to the past. For reasons that are easy to understand, the vertical line is called a contracting fiber, the horizontal line a dilating fiber.

We see now the complete analogy with bifurcation theory. A contracting fiber and a dilating fiber correspond to two realizations of dynamics, each involving symmetry-breaking and appearing in pairs. The contracting fiber corresponds to equilibrium in the far distant past, the dilating fiber to the future. We therefore have two Markov chains oriented in opposite time directions.

Now we have to make the transistion from intrinsically random to intrinsically irreversible systems. To do so we must understand more precisely the difference between contracting and dilating fibers. We have seen that another system as unsta-

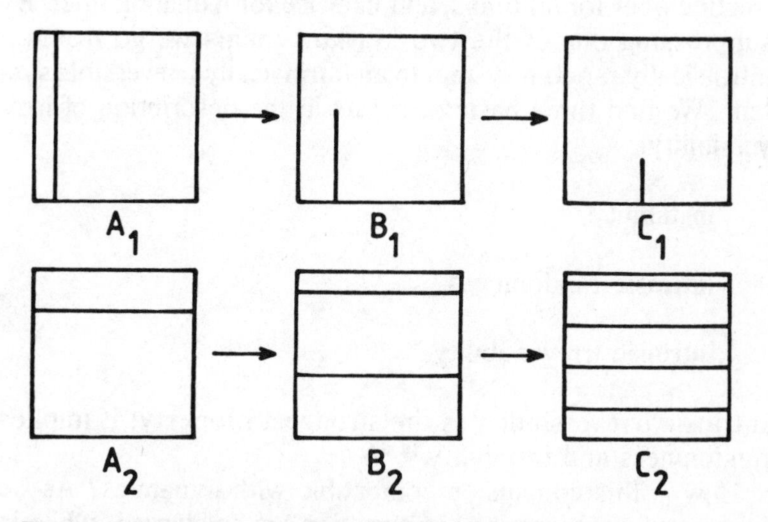

Figure 40. Contracting and dilating fibers in the baker transformation; time going on, the contracting fiber A_1 is shortened (sequence A_1, B_1, C_1), while the dilating fibers are duplicated (sequences A_2, B_2, C_2).

ble as the baker transformation can describe the scattering of hard spheres. Here contracting and dilating fibers have a simple physical interpretation. A contracting fiber corresponds to a collection of hard spheres whose velocities are randomly distributed in the far distant past, and all become parallel in the far distant future. A dilating fiber corresponds to the inverse situation, in which we start with parallel velocities and go to a random distribution of velocities. Therefore the difference is very similar to the one between incoming waves and outgoing waves in the example given by Popper. The exclusion of the contracting fibers corresponds to the experimental fact that whatever the ingenuity of the experimenter, he will never be able to control the system to produce parallel velocities after an arbitrary number of collisions. Once we exclude contracting fibers we are left with only one of the two possible Markov chains we have introduced. In other words, the second law becomes a selection principle of initial conditions. Only initial conditions that go to equilibrium in the *future* are retained.

Obviously the validity of this selection principle is maintained by dynamics. It can easily be seen in the example of the baker transformation that the contracting fiber remains a contracting fiber for all times, and likewise for a dilating fiber. By suppressing one of the two Markov chains we go from an intrinsically random system to an intrinsically irreversible system. We find three basic elements in the description of irreversibility:

instability
↑
intrinsic randomness
↑
intrinsic irreversibility

Intrinsic irreversibility is the strongest property: it implies randomness and instability.[14,15]

How is this conclusion compatible with dynamics? As we have seen, in dynamics "information" is conserved, while in Markov chains information is lost (and entropy therefore increases; see Chapter VIII). There is, however, no contradiction; when we go from the dynamic description of the baker transformation to the thermodynamic description, we have to modify our distribution function; the "objects" in terms of which entropy increases are different from the ones considered in dynamics. The new distribution function, $\hat{\rho}$, corresponds to an intrinsically time-oriented description of the dynamic system. In this book we cannot dwell on the mathematical aspects of this transformation. Let us only emphasize that it has to be *non*canonical (see Chapter II). We must abandon the usual formulations of dynamics to reach the thermodynamic description.

It is quite remarkable that such a transformation exists and that as a result we can now unify dynamics and thermodynamics, the physics of being and the physics of becoming. We shall come back to these new thermodynamic objects later in this chapter as well as in our concluding chapter. Let us emphasize only that at equilibrium, whenever entropy reaches its maximum, these objects must behave randomly.

It also seems quite remarkable that irreversibility emerges, so to speak, from instability, which introduces irreducible statistical features into our description. Indeed, what could an arrow of time mean in a deterministic world in which both future and past are contained in the present? It is because the future is not contained in the present and that we go from the present to the future that the arrow of time is associated with the transition from present to future. This construction of irreversibility out of randomness has, we believe, many consequences that go beyond science proper and to which we shall come back in our concluding chapter. Let us clarify the difference between the states permitted by the second law and those it prohibits.

The Entropy Barrier

Time flows in a single direction, from past to future. We cannot manipulate time, we cannot travel back to the past. Travel through time has preoccupied writers, from *The Thousand and One Nights* to H. G. Wells' *The Time Machine*. In our time, Nabokov's short novel, *Look at the Harlequins!,*[16] describes the torment of a narrator who finds himself as unable to switch from one spatial direction to the other as we are to "twirl time." In the fifth volume of *Science and Civilization in China,* Needham describes the dream of the Chinese alchemists: their supreme aim was not to achieve the transmutation of metals into gold but to manipulate time, to reach immortality through a radical slowdown of natural decaying processes.[17] We are now better able to understand why we cannot "twirl time," to use Nabokov's expression.

An infinite entropy barrier separates possible initial conditions from prohibited ones. Because this barrier is infinite, technological progress never will be able to overcome it. We have to abandon the hope that one day we will be able to travel back into our past. The situation is somewhat analogous to the barrier presented by the velocity of light. Technological progress can bring us closer to the velocity of light, but in the present views of physics we will never cross it.

To understand the origin of this barrier, let us return to the expression of the $\mathcal{H}$ quantity as it appears in the theory of Markov chains (see Chapter VIII). To each distribution we can associate a number—the corresponding value of $\mathcal{H}$. We can say that to each distribution corresponds a well-defined information content. The higher the information content, the more difficult it will be to realize the corresponding state. What we wish to show here is that the initial distribution prohibited by the second law would have an infinite information content. That is the reason why we can neither realize them nor find them in nature.

Let us first come back to the meaning of $\mathcal{H}$ as presented in Chapter VIII. We have to subdivide the relevant phase space into sectors or boxes. With each box k, we associate an probability $P_{eqm}(k)$ at equilibrium as well as a non-equilibrium probability $P(k,t)$.

The $\mathcal{H}$ is a measure of the difference between $P(k,t)$ and $P_{eqm}(k)$, and vanishes at equilibrium when this difference disappears. Therefore, to compare the Baker transformation with Markov chains, we have to make more precise the corresponding choice of boxes. Suppose we consider a system at time 2 (see Figure 39), and suppose that this system originated at time t_i. Then, a result of our dynamical theory is that the boxes correspond to all possible intersections among the partitions between time t_i and $t=2$. If we consider now Figure 39, we see that when t_i is receding towards the past, the boxes will become steadily thinner as we have to introduce more and more vertical subdivisions. This is expressed in Figure 41, sequence B, where, going from top to bottom, we have $t_i = 1, 0, -1$, and finally $t_i = -2$. We see indeed that the number of boxes increases in this way from 4 to 32.

Once we have the boxes, we can compare the non-equilibrium distribution with the equilibrium distribution for each box. In the present case, the non-equilibrium distribution is either a dilating fiber (sequence A) or a contracting fiber (sequence C). The important point to notice is that when t_i is receding to the past, the dilating fiber occupies an increasing large number of boxes: for $t_i = -1$ it occupies 4 boxes, for $t_i = -2$ it occupies 8 boxes, and so on.

As a result, when we apply the formula given in Chapter VIII, we obtain a finite result, even if the number of boxes goes to infinity for $t_i \to -\infty$.

In contrast, the contracting fiber remains always localized in 4 boxes whatever t_i. As a result, $\mathcal{H}$, when applied to a con-

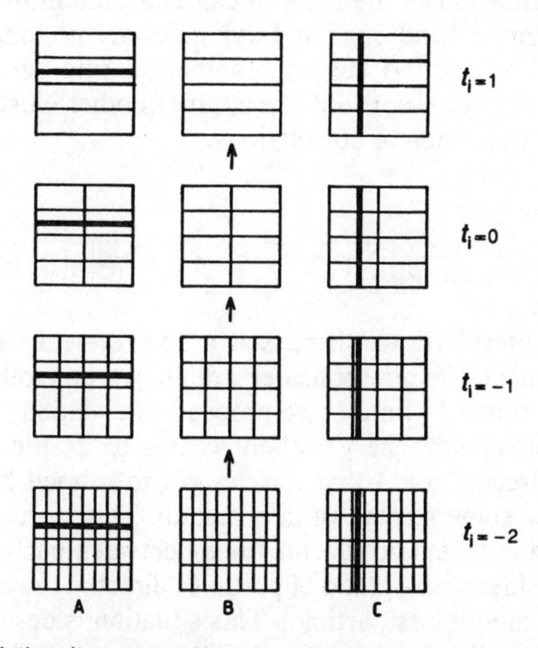

Figure 41. Dilating (sequence A) and contracting (sequence C) fibers cross various numbers of the boxes which subdivide a Baker transformation phase space. All "squares" on a given sequence refer to the same time, $t = 2$, but the number of the boxes subdividing each square depends on the initial time of the system t_i.

tracting fiber, diverges to infinity when t_i recedes to the past. In summary, the difference between a dynamical system and the Markov chain is that the number of boxes to be considered in a dynamical system is infinite. It is this fact that leads to a selection principle. Only measures or probabilities, which in the limit of infinite number of boxes give a finite information or a finite $\mathcal{H}$ quantity, can be prepared or observed. This excludes contracting fibers.[18] For the same reason we must also exclude distributions concentrated on a single point. Initial conditions corresponding to a single point in unstable systems would again correspond to infinite information and are therefore impossible to realize or observe. Again we see that the second law appears as a selection principle.

In the classical scheme, initial conditions were arbitrary. This is no longer so for unstable systems. Here we can associate an information content to each initial condition, and this information content itself depends on the dynamics of the system (as in the baker transformation we used the successive fragmentation of the cells to calculate the information content). Initial conditions and dynamics are no longer independent. The second law as a selection rule seems to us so important that we would like to give another illustration based on the dynamics of correlations.

The Dynamics of Correlations

In Chapter VIII we discussed briefly the velocity inversion experiment. We may consider a dilute gas and follow its evolution in time. At time t_0 we proceed to a velocity inversion of each molecule. The gas then returns to its initial state. We have already noted that for the gas to retrace its past there must be some storage of information. This storage can be described in terms of "correlations" between particles.[19]

Consider first a cloud of particles directed toward a target (a heavy, motionless particle). This situation is described in Figure 42. In the far distant past, there were no correlations between particles. Now, scattering has two effects, already mentioned in Chapter VIII. It disperses the particles (it makes the velocity distribution more symmetrical) and, in addition, it produces correlations between the scattered particles and the scatterer. The correlations can be made explicit by performing a velocity inversion (that is, by introducing a spherical mirror). Figure 43 represents this situation (the wavy lines represent the correlations). Therefore, the role of scattering is as follows: In the direct process, it makes the velocity distribution more symmetrical and creates correlations; in the inverse process, the velocity distribution becomes less symmetrical and correlations disappear. We see that the consideration of cor-

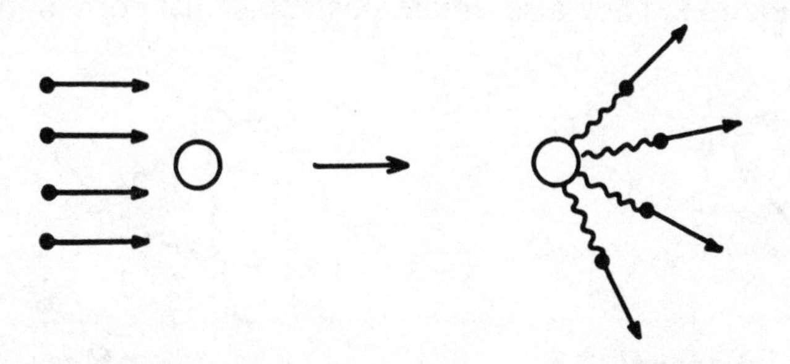

Figure 42. Scattering of particles. Initially all particles have the same velocity. After the collision, the velocities are no more identical, and the scattered particles are correlated with the scatterer (correlations are always represented by wavy lines).

relations introduces a basic distinction between the direct and the inverse processes.

We can apply our conclusions to many-body systems. Here also we may consider two types of situations: in one, uncorrelated particles enter, are scattered, and correlated particles are produced (see Figure 44). In the opposite situation, correlated

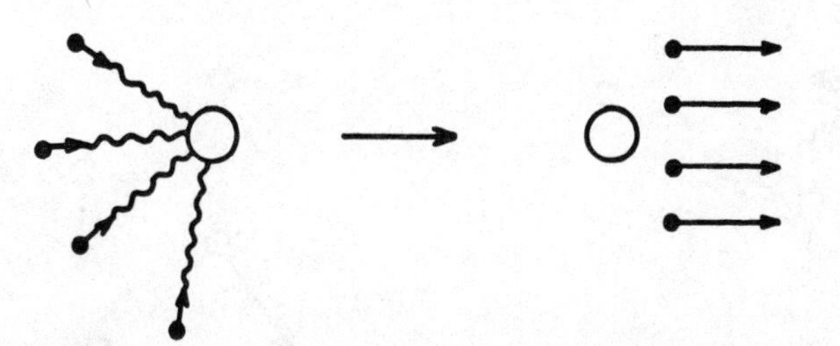

Figure 43. Effect of a velocity inversion after a collision: after the new "inverted" collision, the correlations are suppressed and all particles have the same velocity.

particles enter, the correlations are destroyed through collisions, and uncorrelated particles result (see Figure 45).

The two situations differ in the temporal order of collisions and correlations. In the first case, we have "postcollisional" correlations. With this distinction between pre- and postcollisional correlations in mind, let us return to the velocity inversion experiment. We start at $t=0$, with an initial state corresponding to no correlations between particles. During the time $0 \rightarrow t_0$ we have a "normal" evolution. Collisions bring the velocity distribution closer to the Maxwellian equilibrium distribution. They also create postcollisional correlations be-

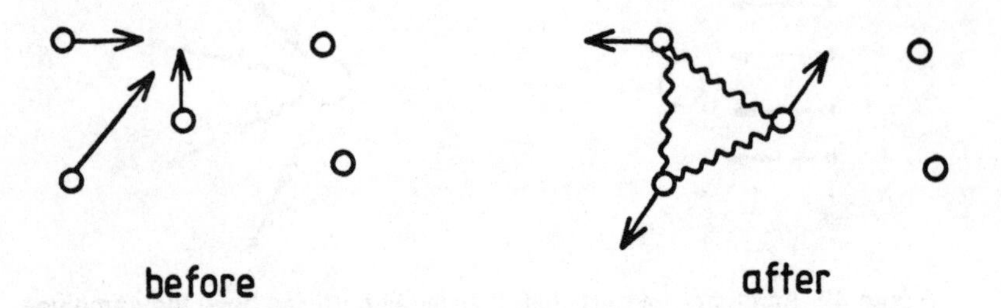

Figure 44. Creation of postcollisional correlations represented by wavy lines; for details see text.

tween the particles. At t_0 after the velocity inversion, a completely new situation arises. Postcollisional correlations are now transformed into precollisional correlations. In the time interval between t_0 and $2t_0$, these precollisional correlations disappear, the velocity distribution becomes less symmetrical, and at time $2t_0$ we are back in the noncorrelational state. The history of this system therefore has two stages. During the

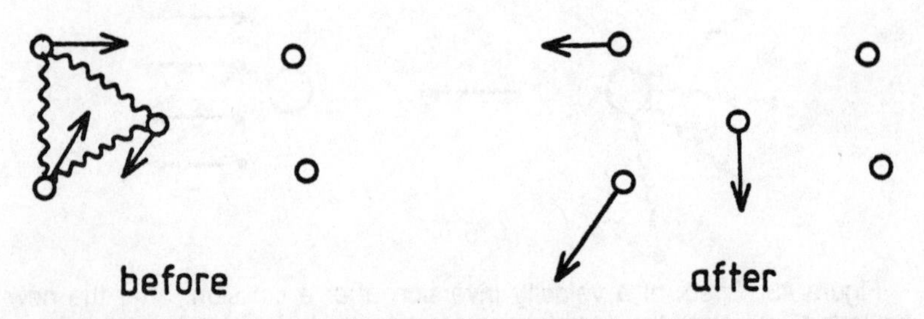

Figure 45. Destruction of precollisional correlations (wavy lines) through collisions.

first, collisions are transformed into correlations; in the second, correlations turn back into collisions. Both types of processes are compatible with the laws of dynamics. Moreover, as we have mentioned in Chapter VIII, the total "information" described by dynamics remains constant. We have also seen that in Boltzmann's description the evolution from time 0 till t_0 corresponds to the usual decrease of $\mathcal{H}$, while from t_0 to $2t_0$ we have an abnormal situation: $\mathcal{H}$ would increase and entropy decrease. We would then be able to devise experiments in the laboratory or on computers in which the second law would be violated! The irreversibility during time $0-t_0$ would be "compensated" by "anti-irreversibility" during time t_0-2t_0.

This is quite unsatisfactory. All these difficulties disappear if we go, as in the baker transformation, to the new "thermodynamic representation" in terms of which dynamics becomes a probabilistic process like a Markov chain. We must also take into account that velocity inversion is not a "natural" process; it requires that "information" be given to molecules from the outside for them to invert their velocity. We need a kind of Maxwellian demon to perform the velocity inversions, and Maxwell's demon has a price. Let us represent the $\mathcal{H}$ quantity (for the probabilistic process) as a function of time. This is done in Figure 46. In this approach, in contrast with Boltzmann's, the effect of correlations is retained in the new definition of $\mathcal{H}$. Therefore at the velocity inversion point t_0 the $\mathcal{H}$ quantity will jump, since we abruptly create abnormal precollisional correlations that will have to be destroyed later. This jump corresponds to the entropy or information price we have to pay.

Now we have a faithful representation of the second law: at *every moment* the $\mathcal{H}$ quantity decreases (or the entropy increases). There is one exception at time t_0: $\mathcal{H}$ jumps upward, but that corresponds to the very moment at which the system is open. We can invert the velocities only by acting from the outside.

There is another essential point: at time t_0 the new $\mathcal{H}$ quantity has two different values, one for the system before velocity inversion and the other after a velocity inversion. These two situations have different entropies. This resembles what occurs in the baker transformation when the contracting and dilating fibers are velocity inversions of each other.

Suppose we wait a sufficient time before making the velocity inversion. The postcollisional correlations would have an arbitrary range, and the entropy price for velocity inversion would become too high. The velocity inversion would then require too high an entropy price and thus would be excluded. In physical terms this means that the second law excludes persistent long-range precollisional correlations.

The analogy with the macroscopic description of the second law is striking. From the point of view of energy conservation (see Chapters IV and V), heat and work play the same role,

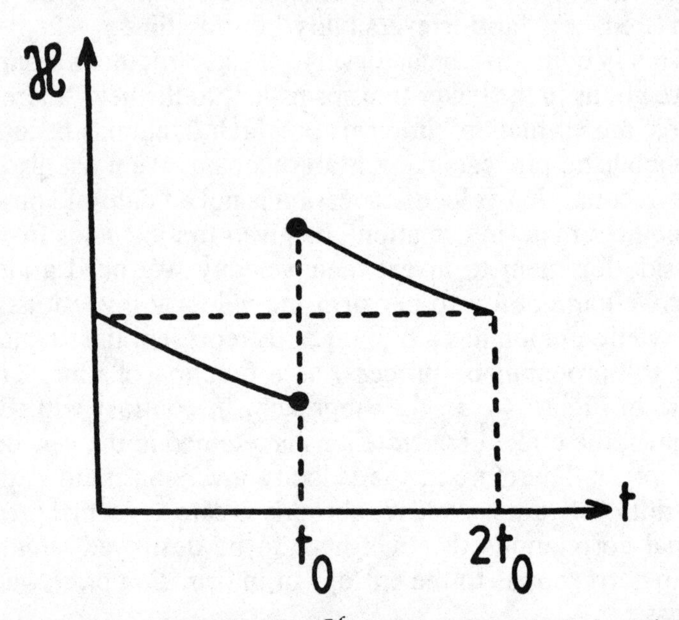

Figure 46. Time variation of the $\mathcal{H}$-function in the velocity inversion experiment: at time t_0, the velocities are inversed and $\mathcal{H}$ presents a discontinuity. At time $2t_0$ the system is in the same state as at time 0, and $\mathcal{H}$ recovers the value it had initially. At all times (except at t_0), $\mathcal{H}$ is decreasing. The important fact is that at time t_0 the $\mathcal{H}$-quantity takes two different values (see text).

but no longer from the point of view of the second law. Briefly speaking, work is a more coherent form of energy and always can be converted into heat, but the inverse is not true. There is on the microscopic level a similar distinction between collisions and correlations. From the point of view of dynamics, collisions and correlations play equivalent roles. Collisions give rise to correlations, and correlations may destroy the effect of collisions. But there is an essential difference. We can control collisions and produce correlations, but we cannot control correlations in a way that will destroy the effects collisions have brought into the system. It is this essential difference that is missing in dynamics but that can be incorporated into thermodynamics. Note that thermodynamics does not enter into conflict with dynamics at any point. It adds an additional, essential element to our understanding of the physical world.

Entropy as a Selection Principle

It is amazing how closely the microscopic theory of irreversible processes resembles traditional macroscopic theory. In both cases entropy initially has a negative meaning. In its macroscopic aspect it prohibits some processes, such as heat flowing from cold to hot. In its microscopic aspect it prohibits certain classes of initial conditions. The distinction between what is permitted and what is prohibited is maintained in time by the laws of dynamics. It is from the negative aspect that the positive aspect emerges: the existence of entropy together with its probability interpretation. Irreversibility no longer emerges as if by a miracle at some macroscopic level. Macroscopic irreversibility only makes apparent the time-oriented polarized nature of the universe in which we live.

As we have emphasized repeatedly, there exist in nature systems that behave reversibly and that may be fully described by the laws of classical or quantum mechanics. But most systems of interest to us, including all chemical systems and therefore all biological systems, are time-oriented on the macroscopic level. Far from being an "illusion," this expresses a broken time-symmetry on the microscopic level. Irreversibility is either true on *all* levels or on none. It cannot emerge as if by a miracle, by going from one level to another.

Also we have already noticed, irreversibility is the starting point of other symmetry breakings. For example, it is generally accepted that the difference between particles and antiparticles could arise only in a nonequilibrium world. This may be extended to many other situations. It is likely that irreversibility also played a role in the appearance of chiral symmetry through the selection of the appropriate bifurcation. One of the most active subjects of research now is the way in which irreversibility can be "inscribed" into the structure of matter.

The reader may have noticed that in the derivation of microscopic irreversibility we have been concentrating on classical dynamics. However, the ideas of correlations and the distinction between pre- and postcollisional correlations apply to quantum systems as well. The study of quantum systems is more complicated than the study of classical systems. There is a reason for this: the difference between classical and quantum mechanics. Even small classical systems, such as those formed by a few hard spheres, may present intrinsic irreversibility. However, to reach irreversibility in quantum systems we need large systems, such as those realized in liquids, gases, or in field theory. The study of large systems is obviously more difficult mathematically, and that is why we will not go into the matter here. However, the situation remains essentially the same in quantum theory. There also irreversibility begins as the result of the limitation of the concept of wave function due to a form of quantum instability.

Moreover, the idea of collisions and correlations may also be used in quantum theory. Therefore, as in classical theory, the second law prohibits long-range precollisional correlations.

The transition to a probability process introduces new entities, and it is in terms of *these new entities that the second law can be understood as an evolution from order to disorder.* This is an important conclusion. The second law leads to a new concept of matter. We would like to describe this concept now.

Active Matter

Once we associate entropy with a dynamic system, we come back to Boltzmann's conception: the probability will be maximum at equilibrium. The units we use to describe thermodynamic evolution will therefore behave in a chaotic way at equilibrium. In contrast, in near-equilibrium conditions correlations and coherence will appear.

We come to one of our main conclusions: At all levels, be it the level of macroscopic physics, the level of fluctuations, or the microscopic level, *nonequilibrium is the source of order. Nonequilibrium brings "order out of chaos."* But as we already mentioned, the concept of order (or disorder) is more complex than was thought. It is only in some limiting situations, such as with dilute gases, that it acquires a simple meaning in agreement with Boltzmann's pioneering work.

Let us once more contrast the dynamic description of the physical world in terms of forces and fields with the thermodynamic description. As we mentioned, we can construct computer experiments in which interacting particles initially distributed at random form a lattice. The dynamic interpretation would be the appearance of order through interparticle forces. The thermodynamic interpretation is, on the contrary, the approach to *disorder* (when the system is isolated), but to disorder expressed in quite different units, which are in this case collective modes involving a large number of particles. It seems to us worthwhile to reintroduce the neologism we used in Chapter VI to define the new units in terms of which the system is incoherent at equilibrium: we call them "hypnons," sleepwalkers, since they ignore each other at equilibrium. Each of them may be as complex as you wish (think about molecules of the complexity of an enzyme), but at equilibrium their complexity is turned "inward." Again, inside a molecule there is an intensive electric field, but this field in a dilute gas is negligible as far as other molecules are concerned.

One of the main subjects in present-day physics is the problem of elementary particles. However, we know that elementary particles are far from elementary. New layers of structure are disclosed at higher and higher energies. But what, after all, is an elementary particle? Is the planet earth an elementary particle? Certainly not, because part of this energy is in its interaction with the sun, the moon, and the other planets. The concept of elementary particles requires an "autonomy" that is very difficult to describe in terms of the usual concepts. Take the case of electrons and photons. We are faced with a dilemma: either there are no well-defined particles (because the energy is partly between the electrons and protons), or there are noninteracting particles if we can eliminate the interaction. Even if we knew how to do that, it seems too radical a procedure. Electrons absorb photons or emit photons. A way out may be to go to the physics of processes. The units, the elementary particles, would then be defined as hypnons, as the entities that evolve independently at equilibrium. We hope that there soon will be experiments available to test this hypothesis; it would be quite appealing if atoms interacting with photons (or unstable elementary particles) already carried the arrow of time that expresses the global evolution of nature.

A subject widely discussed today is the problem of cosmic evolution. How could the world near the moment of the Big Bang be so "ordered"? Yet this order is necessary if we wish to understand cosmic evolution as the gradual movement from order to disorder.

To give a satisfactory answer we need to know what "hypnons" could have been appropriate to the extreme conditions of temperature and density that characterized the early universe. Thermodynamics alone will not, of course, solve these problems; neither will dynamics, even in its most refined form field theory. That is why the unification of dynamics and thermodynamics opens new perspectives.

In any case, it is striking how the situation has changed since the formulation of the second law of thermodynamics one hundred fifty years ago. At first it seemed that the atomistic view contradicted the concept of entropy. Boltzmann attempted to save the mechanistic world view at the cost of reducing the second law to a statement of probability with great practical importance but no fundamental significance. We do not know what the definitive solution will be; but today the situation is radically different. Matter is not given. In the present-day view it has to be constructed out of a more fundamental concept in terms of quantum fields. In this construction of matter, thermodynamic concepts (irreversibility, entropy) have a role to play.

Let us summarize what has been achieved here. The central role of the second law (and of the correlative concept of irreversibility) at the level of macroscopic systems has already been emphasized in Books One and Two.

What we have tried to show in Book Three is that we now can go beyond the macroscopic level, and discover the microscopic meaning of irreversibility.

However, this requires basic changes in the way in which we

conceive the fundamental laws of physics. It is only when the classical point of view is lost—as it is in the case of sufficiently unstable systems—that we can speak of 'intrinsic randomness' and 'intrinsic irreversibility.'

It is for such systems that we may introduce a new enlarged description of time in terms of the operator time *T*. As we have shown in the example of the Baker transformation (Chapter IX "From randomness to irreversibility"), this operator has as eigenfunctions partitions of the phase space (see Figure 39).

We come therefore to a situation quite reminiscent of that of quantum mechanics. We have indeed two possible descriptions. Either we give ourselves a point in phase space, and then we don't know to which partition it belongs, and therefore we don't know its internal age; or we know its internal age, but then we know only the partition, but not the exact localization of the point.

Once we have introduced the internal time *T*, we can use entropy as a selection principle to go from the initial description in terms of the distribution function ρ to a new one, $\hat{\rho}$ where the distribution $\hat{\rho}$ has an intrinsic arrow of time, satisfying the second law of thermodynamics. The basic difference between ρ and $\hat{\rho}$ appears when these functions are expanded in terms of the eigenfunction of the operator time *T* (see Chapter VII, "The rise of quantum mechanics"). In ρ, all internal ages, be they from past or future, appear symmetrically. In contrast, in $\hat{\rho}$, past and future play different roles. The past is included, but the future remains uncertain. That is the meaning of the arrow of time. The fascinating aspect is that there appears now a relation between initial conditions and the laws of change. A state with an arrow of time emerges from a law, which has also an arrow of time, and which transforms the state, however keeping this arrow of time.

We have concentrated mostly on the classical situation.[20] However, our analysis applies as well to quantum mechanics, where the situation is more complicated, as the existence of Planck's constant destroys already the concept of a trajectory, and leads therefore also to a kind of delocalization in phase space. In quantum mechanics we have therefore to superpose the quantum delocalization with delocalization due to irreversibility.

As emphasized in Chapter VII, the two great revolutions in

the physics of our century correspond to the incorporation, in the fundamental structure of physics, of impossibilities foreign to classical mechanics: the impossibility of signals propagating with a velocity larger than the velocity of light, and the impossibility of measuring simultaneously coordinates and momentum.

It is not astonishing that the second principle, which as well limits our ability to manipulate matter, also leads to deep changes in the structure of the basic laws of physics.

Let us conclude this part of our monograph with a word of caution. The phenomenological theory of irreversible processes is at present well established. In contrast, the basic microscopic theory of irreversible processes is quite new. At the time of correcting the proofs of this book, experiments are in preparation to test these views. As long as they have not been performed, a speculative element is unavoidable.

CONCLUSION

FROM EARTH TO HEAVEN—THE REENCHANTMENT OF NATURE

In any attempt to bridge the domains of experience belonging to the spiritual and physical sides of our nature, time occupies the key position.

A. S. EDDINGTON[1]

An Open Science

Science certainly involves manipulating nature, but it is also an attempt to understand it, to dig deeper into questions that have been asked generation after generation. One of these questions runs like a leitmotiv, almost as an obsession, through this book, as it does through the history of science and philosophy. This is the question of the relation between being and becoming, between permanence and change.

We have mentioned pre-Socratic speculations: Is change, whereby things are born and die, imposed from the outside on some kind of inert matter? Or is it the result of the intrinsic and independent activity of matter? Is an external driving force necessary, or is becoming inherent in matter? Seventeenth-century science arose in opposition to the biological model of a spontaneous and autonomous organization of natural beings. But it was confronted with another fundamental alternative. Is nature intrinsically random? Is ordered behavior merely the transient result of the chance collisions of atoms and of their unstable associations?

One of the main sources of fascination in modern science was precisely the feeling that it had discovered eternal laws at the core of nature's transformations and thus had exorcised time and becoming. This discovery of an order in nature produced the feeling of intellectual security described by French sociologist Lévy-Bruhl:

> Our feeling of intellectual security is so deeply anchored in us that we even do not see how it could be shaken. Even if we suppose that we could observe some phenomenon seemingly quite mysterious, we still would remain persuaded that our ignorance is only provisional, that this phenomenon must satisfy the general laws of causality, and that the reasons for which it has appeared will be determined sooner or later. *Nature around us is order and reason, exactly as is the human mind.* Our everyday activity implies a perfect confidence in the universality of the laws of nature.[2]

This feeling of confidence in the "reason" of nature has been shattered, partly as the result of the tumultuous growth of science in our time. As we stated in the Preface, our vision of nature is undergoing a radical change toward the multiple, the temporal, and the complex. Some of these changes have been described in this book.

We were seeking general, all-embracing schemes that could be expressed in terms of eternal laws, but we have found time, events, evolving particles. We were also searching for symmetry, and here also we were surprised, since we discovered symmetry-breaking processes on all levels, from elementary particles up to biology and ecology. We have described in this book the clash between dynamics, with the temporal symmetry it implies, and the second law of thermodynamics, with its directed time.

A new unity is emerging: irreversibility is a source of order at all levels. Irreversibility is the mechanism that brings order out of chaos. How could such a radical transformation of our views on nature occur in the relatively short time span of the past few decades? We believe that it shows the important role intellectual construction plays in our concept of reality. This was very well expressed by Bohr, when he said to Werner Heisenberg on the occasion of a visit at Kronberg Castle:

> Isn't it strange how this castle changes as soon as one imagines that Hamlet lived here? As scientists we believe that a castle consists only of stones, and admire the way the architect put them together. The stones, the green roof with its patina, the wood carvings in the church, constitute the whole castle. None of this should be changed by the fact that Hamlet lived here, and yet it is changed completely. Suddenly the walls and the ramparts speak a different language. . . . Yet all we really know about Hamlet is that his name appears in a thirteenth-century chronicle. . . . But everyone knows the questions Shakespeare had him ask, the human depths he was made to reveal, and so he too had to have a place on earth, here in Kronberg.[3]

The question of the meaning of reality was the central subject of a fascinating dialogue between Einstein and Tagore.[4] Einstein emphasized that science had to be independent of the existence of any observer. This led him to deny the reality of time as irreversibility, as evolution. On the contrary, Tagore maintained that even if absolute truth could exist, it would be inaccessible to the human mind. Curiously enough, the present evolution of science is running in the direction stated by the great Indian poet. Whatever we call reality, it is revealed to us only through the active construction in which we participate. As it is concisely expressed by D. S. Kothari, "The simple fact is that no measurement, no experiment or observation is possible without a relevant theoretical framework."[5]

Time and Times

The statement that time is basically a geometric parameter that makes it possible to follow the unfolding of the succession of dynamical states has been asserted in physics for more than three centuries. Emile Meyerson[6] tried to describe the history of modern science as the gradual implementation of what he regarded as a basic category of human reason: the different and the changing had to be reduced to the identical and the permanent. Time had to be eliminated.

Nearer to our own time, Einstein appears as the incarnation of this drive toward a formulation of physics in which no reference to irreversibility would be made on the fundamental level.

An historic scene took place at the Société de Philosophie in Paris on April 6, 1922,[7] when Henri Bergson attempted to defend the cause of the multiplicity of coexisting "lived" times against Einstein. Einstein's reply was absolute: he categorically rejected "philosophers' time." Lived experience cannot save what has been denied by science.

Einstein's reaction was somewhat justified. Bergson had obviously misunderstood Einstein's theory of relativity. However, there also was some prejudice on Einstein's part: *durée*, Bergson's "lived time," refers to the basic dimensions of becoming, the irreversibility that Einstein was willing to admit only at the phenomenological level. We have already referred to Einstein's conversations with Carnap (see Chapter VII). For him distinctions among past, present, and future were outside the scope of physics.

It is fascinating to follow the correspondence between Einstein and the closest friend of his young days in Zurich, Michele Besso.[8] Although he was an engineer and scientist, toward the end of his life Besso became increasingly concerned with philosophy, literature, and the problems that surround the core of human existence. Untiringly he kept asking the same questions: What is irreversibility? What is its relationship with the laws of physics? And untiringly Einstein would answer with a patience he showed only to this closest friend: irreversibility is merely an illusion produced by "improbable" initial conditions. This dialogue continued over many years until Besso, older than Einstein by eight years, passed away, only a few months before Einstein's death. In a last letter to Besso's sister and son, Einstein wrote: "Michele has left this strange world just before me. This is of no importance. For us convinced physicists the distinction between past, present and future is an illusion, although a persistent one." In Einstein's drive to perceive the basic laws of physics, the intelligible was identified with the immutable.

Why was Einstein so strongly opposed to the introduction of irreversibility into physics? We can only guess. Einstein was a rather lonely man; he had few friends, few coworkers, few students. He lived in a sad time: the two World Wars, the rise of anti-Semitism. It is not surprising that for Einstein science was the road that led to victory over the turmoil of time. What a contrast, however, with his scientific work. His world was full of observers, of scientists situated in various coordinate systems in motion with one another, situated on various stars differing by their gravitational fields. All these observers were exchanging information through signals all over the universe. What Einstein wanted to preserve above all was the objective meaning of this communication. However, we can perhaps state that Einstein stopped short of accepting that communication and irreversibility are closely related. Communication is at the base of what probably is the most irreversible process accessible to the human mind, the progressive increase of knowledge.

The Entropy Barrier

In Chapter IX we described the second law as a selection principle: to each initial condition there corresponds an "information." All initial conditions for which this information is finite are permitted. However, to reverse the direction of time we would need infinite information; we cannot produce situations that would evolve into our past! This is the entropy barrier we have introduced.

There is an interesting analogy with the concept of the velocity of light as the maximum velocity of transmission of signals. As we have seen in Chapter VII, this is one of the basic postulates of Einstein's relativity theory. The existence of the velocity of light barrier is necessary to give meaning to causality. Suppose we could, in a science-fiction ship, leave the earth at a velocity greater than the velocity of light. We could overtake light signals and in this way precede our own past. Likewise, the entropy barrier is necessary to give meaning to communication. We have already mentioned that irreversibility and communication are closely related. Norbert Wiener has argued that the existence of two directions of time would have disastrous consequences. It is worthwhile to cite a passage from his well-known book *Cybernetics:*

> Indeed, it is a very interesting intellectual experiment to make the fantasy of an intelligent being whose time should run the other way to our own. To such a being, all communication with us would be impossible. Any signal he might send would reach us with a logical stream of consequents from his point of view, antecedents from ours. These antecedents would already be in our experience, and would have served to us as the natural explanation of his signal, without presupposing an intelligent being to have sent it. If he drew us a square, we should see the remains of his figure as its precursors, and it would seem to be the curious crystallization—always perfectly explainable—of these remains. Its meaning would seem to be as fortuitous as the faces we read into mountains and cliffs. The drawing of the square would appear to us as a catastrophe—sudden indeed, but explainable by natural laws—by which that square would cease to exist. Our counterpart would have exactly similar ideas concerning us. *Within any world with which we can communicate, the direction of time is uniform.*[9]

It is precisely the infinite entropy barrier that guarantees the uniqueness of the direction of time, the impossibility of switching from one direction of time to the opposite one.

In the course of this book, we have stressed the importance of demonstrations of impossibility. In fact, Einstein was the first to grasp that importance when he based his concept of relative simultaneity on the impossibility of transmitting information at a velocity greater than that of light. The whole theory of relativity is built around the exclusion of "unobservable" simultaneities. Einstein considered this step as somewhat similar to the step taken in thermodynamics when perpetual motion was excluded. But some of his contemporaries—Heisenberg, for example—emphasized an important difference between these two impossibilities. In the case of thermodynamics, a certain *situation* is defined as being absent from nature; in the case of relativity, it is an *observation* that is defined as impossible—that is, a type of dialogue, of communication between nature and the person who describes it. Thus Heisenberg saw himself as following Einstein's example, in spite of Einstein's skepticism, when he grounded quantum mechanics on the exclusion of what the quantum uncertainty principle defines as unobservable.

As long as the second law was considered to express only a practical improbability, it had little theoretical interest. You could always hope to overcome it by sufficient technical skill. But we have seen that this is not so. At its roots there is a selection of possible initial states. It is only after these states have been selected that the probability interpretation becomes possible. Indeed, as Boltzmann stated for the first time, the increase of entropy expresses the increase of probability, of disorder. However, his interpretation results from the conclusion that entropy is a selection principle breaking the time symmetry. It is only after this symmetry-breaking that any probabilistic interpretation becomes possible.

In spite of the fact that we have recouped much of Boltzmann's interpretation of entropy, the basis of our interpretation of his second law is radically different, since we have in succession

the second law as a symmetry-breaking selection principle
↓
probabilistic interpretation
↓
irreversibility as increase of disorder

It is only the unification of dynamics and thermodynamics through the introduction of a new selection principle that gives the second law its fundamental importance as the evolutionary paradigm of the sciences. This point is of such importance that we shall dwell on it in more detail.

The Evolutionary Paradigm

The world of dynamics, be it classical or quantum, is a reversible world. As we have emphasized in Chapter VIII, no evolution can be ascribed to this world; the "information" expressed in terms of dynamical units remains constant. It is therefore of great importance that the existence of an evolutionary paradigm can now be established in physics—not only on the level of macroscopic description but also on all levels. Of course, there are conditions: as we have seen, a minimum complexity is necessary. But the immense importance of irreversible processes shows that this requirement is satisfied for most systems of interest. Remarkably, the perception of oriented time increases as the level of biological organization increases and probably reaches its culminating point in human consciousness.

How general is this evolutionary paradigm? It includes isolated systems that evolve to disorder and open systems that evolve to higher and higher forms of complexity. It is not surprising that the entropy metaphor has tempted a number of writers dealing with social or economic problems. Obviously here we have to be careful; human beings are not dynamic objects, and the transition to thermodynamics cannot be formulated as a selection principle maintained by dynamics. On the human level irreversibility is a more fundamental concept, which is for us inseparable from the meaning of our existence. Still it is essential that in this perspective we no longer see the internal feeling of irreversibility as a subjective impression that alienates us from the outside world, but as marking our participation in a world dominated by an evolutionary paradigm.

Cosmological problems are notoriously difficult. We still do not know what the role of gravitation was in the early universe. Can gravitation be included in some form of the second law, or is there a kind of dialectical balance between thermodynamics and gravitation? Certainly irreversibility could not have appeared abruptly in a time-reversible world. The origin of irreversibility is a cosmological problem and would require an analysis of the universe in its first stages. Here our aim is more modest. What does irreversibility mean today? How does it relate to our position in the world we describe?

Actors and Spectators

The denial of becoming by physics created deep rifts within science and estranged science from philosophy. What had originally been a daring wager with the dominant Aristotelian tradition gradually became a dogmatic assertion directed against all those (chemists, biologists, physicians) for whom a qualitative diversity existed in nature. At the end of the nineteenth century this conflict had shifted from inside science to the relation between "science" and the rest of culture, especially philosophy. We have described in Chapter III this aspect of the history of Western thought, with its continual struggle to achieve a new unity of knowledge. The "lived time" of the phenomenologists, the *Lebenswelt* opposed to the objective world of science, may be related to the need to erect bulwarks against the invasion of science.

Today we believe that the epoch of certainties and absolute oppositions is over. Physicists have no privilege whatsoever to any kind of extraterritoriality. As scientists they belong to their culture, to which, in their turn, they make an essential contribution. We have reached a situation close to the one recognized long ago in sociology: Merleau-Ponty had already stressed the need to keep in mind what he termed a "truth within situations."

> So long as I keep before me the ideal of an absolute observer, of knowledge in the absence of any viewpoint, I can only see my situation as being a source of error. But once I have acknowledged that through it I am geared to all actions and all knowledge that are meaningful to me, and that it is gradually filled with everything that may *be* for me, then my contact with the social in the finitude of my situation is revealed to me as the starting point of all truth, including that of science and, since we have some idea of the truth, since we are inside truth and cannot get outside it, all that I can do is define a truth within the situation.[10]

It is this conception of knowledge as both objective and participatory that we have explored through this book.

In his *Themes*[11] Merleau-Ponty also asserted that the "philosophic" discoveries of science, its basic conceptual transformations, are often the result of *negative discoveries,* which provide the occasion and the starting point for a reversal of point of view. Demonstrations of impossibility, whether in relativity, quantum mechanics, or thermodynamics, have shown us that nature cannot be described "from the outside," as if by a spectator. Description is dialogue, communication, and this communication is subject to constraints that demonstrate that we are macroscopic beings embedded in the physical world.

We may summarize the situation as we see it today in the following diagram:

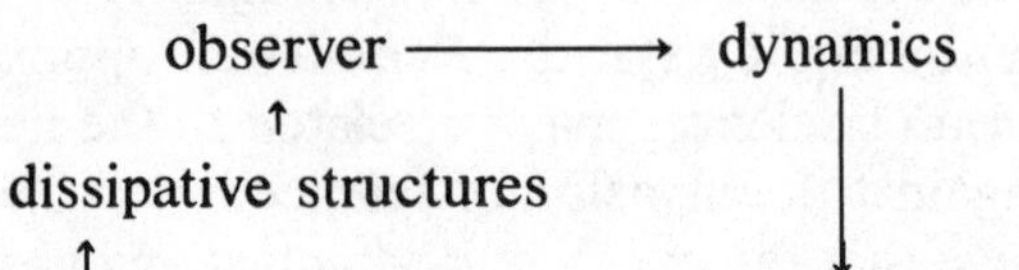

We start from the observer, who measures coordinates and momenta and studies their change in time. This leads him to the discovery of unstable dynamic systems and other concepts of intrinsic randomness and intrinsic irreversibility, as we discussed them in Chapter IX. Once we have intrinsic irreversibility and entropy, we come in far-from-equilibrium systems to dissipative structures, and we can understand the time-oriented activity of the observer.

There is no scientific activity that is not time-oriented. The preparation of an experiment calls for a distinction between "before" and "after." It is only because we are aware of irreversibility that we can recognize reversible motion. Our diagram shows that we have now come full circle, that we can see ourselves as part of the universe we describe.

The scheme we have presented is not an a priori scheme—deducible from some logical structure. There is, indeed, no logical necessity for dissipative structures actually to exist in nature; the "cosmological fact" of a universe far from equilibrium is needed for the macroscopic world to be a world inhabited by "observers"—that is, to be a living world. Our scheme thus does not correspond to a logical or epistemological truth but refers to our condition as macroscopic beings in a world far from equilibrium. Moreover, an essential characteristic of our scheme is that it does not suppose any fundamental mode of description; each level of description is implied by another and implies the other. We need a multiplicity of levels that are all connected, none of which may have a claim to preeminence.

We have already emphasized that irreversibility is not a universal phenomenon. We can perform experiments corresponding to thermodynamic equilibrium in limited portions of space. Moreover, the importance of time scales varies. A stone evolves according to the time scale of geological evolution; human societies, especially in our time, obviously have a much shorter time scale. We have already mentioned that irreversibility starts with a minimum complexity of the dynamic systems involved. It is interesting that with an increase of complexity, going from the stone to human societies, the role of the arrow of time, of evolutionary rhythms, increases. Molecular biology shows that everything in a cell is not alive in the same way. Some processes reach equilibrium, others are dominated by regulatory enzymes far from equilibrium. Similarly, the arrow of time plays quite different roles in the universe around us. From this point of view, in the sense of this time-oriented activity, the human condition seems unique. It seems to us, as we said in Chapter IX, quite important that irreversibility, the arrow of time, implies randomness. "Time is construction." This conclusion, one that Valéry[12] reached quite independently, carries a message that goes beyond science proper.

A Whirlwind in a Turbulent Nature

In our society, with its wide spectrum of cognitive techniques, science occupies a peculiar position, that of a poetical interrogation of nature, in the etymological sense that the poet is a "maker"—active, manipulating, and exploring. Moreover, science is now capable of respecting the nature it investigates. Out of the dialogue with nature initiated by classical science, with its view of nature as an automaton, has grown a quite different view in which the activity of questioning nature is part of its intrinsic activity.

As we have written at the start of this chapter, our feeling of intellectual security has been shattered. We can now appreciate in a nonpolemical fashion the relation between science and philosophy. We have already mentioned the Einstein-Bergson conflict. Bergson was certainly "wrong" on some technical points, but his task as a philosopher was to attempt to make explicit inside physics the aspects of time he thought science was neglecting.

Exploring the implications and the coherence of those fundamental concepts, which appear both scientific and philosophical, may be risky, but it can be very fruitful in the dialogue between science and philosophy. Let us illustrate this with some brief references to Leibniz, Peirce, Whitehead, and Lucretius.

Leibniz introduced the strange concept of monads, noncommunicating physical entities that have "no windows through which something can get in or out." His views have often been dismissed as mad, and still, as we have seen in Chapter II, it is an essential property of all integrable systems that there exist a transformation that may be described in terms of noninteracting entities. These entities translate their own initial state throughout their motion, but at the same time, like monads, they coexist with all the others in a "preestablished" harmony: in this representation, the state of each entity, although perfectly self-determined, reflects the state of the whole system down to the smallest detail.

All integrable systems thus can be viewed as "monadic" systems. Conversely, Leibnizian monadology can be translated into dynamic language: the universe is an integrable system.[13] Monadology thus becomes the most consequential formulation of a universe from which all becoming is eliminated. By considering Leibniz's efforts to understand the activity of matter, we can measure the gap that separates the seventeenth century from our time. The tools were not yet ready; it was impossible, on the basis of a purely mechanical universe, for Leibniz to give an account of the activity of matter. Still some of his ideas, that substance is activity, that the universe is an interrelated unit, remain with us and are today taking on a new form.

We regret that we cannot devote sufficient space to the work of Charles S. Peirce. At least let us cite one remarkable passage:

> You have all heard of the dissipation of energy. It is found that in all transformations of energy a part is converted into heat and heat is always tending to equalize its tem-

> perature. The consequence is that the energy of the universe is tending by virtue of its necessary laws toward a death of the universe in which there shall be no force but heat and the temperature everywhere the same. . . .
>
> But although no force can counteract this tendency, chance may and will have the opposite influence. Force is in the long run dissipative; chance is in the long run concentrative. The dissipation of energy by the regular laws of nature is by these very laws accompanied by circumstances more and more favorable to its reconcentration by chance. There must therefore be a point at which the two tendencies are balanced and that is no doubt the actual condition of the whole universe at the present time.[14]

Once again, Peirce's metaphysics was considered as one more example of a philosophy alienated from reality. But, in fact, today Peirce's work appears to be a pioneering step toward the understanding of the pluralism involved in physical laws.

Whitehead's philosophy takes us to the other end of the spectrum. For him, being is inseparable from becoming. Whitehead wrote: "The elucidation of the meaning of the sentence 'everything flows' is one of metaphysics' main tasks."[15] Physics and metaphysics are indeed coming together today in a conception of the world in which process, becoming, is taken as a primary constituent of physical existence and where, unlike Leibniz' monads, existing entities can interact and therefore also be born and die.

The ordered world of classical physics, or a monadic theory of parallel changes, resembles the equally parallel, ordered, and eternal fall of Lucretius' atoms through infinite space. We have already mentioned the clinamen and the instability of laminar flows. But we can go farther. As Serres[16] points out, the infinite fall provides a *model* on which to base our conception of the natural genesis of the disturbance that causes things to be born. If the vertical fall were not disturbed "without reason" by the clinamen, which leads to encounters and associations between uniformly falling atoms, no nature could be created; all that would be reproduced would be the repetitive connection between equivalent causes and effects governed by the laws of fate *(foedera fati)*.

Denique si semper motus conectitur omnis
et uetere exoritur [*semper*] *novus ordine certo*
nec declinando faciunt primordia motus
principium quoddam quod fati foedera rumpat,
ex infinito ne causam causa sequatur,
libera per terras unde haec animantibus exstat . . . ?[17]

Lucretius may be said to have *invented* the clinamen in the same way that archaeological remains are "invented": one "guesses" they are there before one begins to dig. If only uniformly reversible trajectories existed, where would the irreversible processes we produce and experience come from? The point where the trajectories cease to be determined, where the *foedera fati* governing the ordered and monotonous world of deterministic change break down, marks the beginning of nature. It also marks the beginning of a new science that describes the birth, proliferation, and death of natural beings. "The physics of falling, of repetition, of rigorous concatenation is replaced by the creative science of change and circumstances."[18] The *foedera fati* are replaced by the *foedera naturae,* which, as Serres emphasizes, denote both "laws" of nature—local, singular, historical relations—and an "alliance," a form of contract with nature.

In Lucretian physics we thus again find the link we have discovered in modern knowledge between the choices underlying a physical description and a philosophic, ethical, or religious conception relating to man's situation in nature. The physics of universal connections is set against another science that in the name of law and domination no longer struggles with disturbance or randomness. Classical science from Archimedes to Clausius was opposed to the science of turbulence and of bifurcating changes.

> It is here that Greek wisdom reaches one of its pinnacles. Where man is in the world, of the world, in matter, of matter, he is not a stranger, but a friend, a member of the family, and an equal. He has made a pact with things. Conversely, many other systems and many other sciences are based on breaking this pact. Man is a stranger to the world, to the dawn, to the sky, to things. He hates

them, and fights them. His environment is a dangerous enemy to be fought, to be kept enslaved. . . . Epicurus and Lucretius live in a reconciled universe. Where the science of things and the science of man coincide. I am a disturbance, a whirlwind in turbulent nature.[19]

Beyond Tautology

The world of classical science was a world in which the only events that could occur were those deducible from the instantaneous state of the system. Curiously, this conception, which we have traced back to Galileo and Newton, was not new at that time. Indeed, it can be identified with Aristotle's conception of a divine and immutable heaven. In Aristotle's opinion, it was only the heavenly world to which we could hope to apply an exact mathematical description. In the Introduction, we echoed the complaint that science has "disenchanted" the world. But this disenchantment is paradoxically due to the glorification of the earthly world, henceforth worthy of the kind of intellectual pursuit Aristotle reserved for heaven. Classical science denied becoming, natural diversity, both considered by Aristotle as attributes of the sublunar, inferior world. In this sense, classical science brought heaven to earth. However, this apparently was not the intention of the fathers of modern science. In challenging Aristotle's claim that mathematics ends where nature begins, they did not seek to discover the immutable concealed behind the changing, but rather to extend changing, corruptible nature to the boundaries of the universe. In his *Dialogue Concerning the Two Chief World Systems,* Galileo is amazed at the notion that the world would be a nobler place if the great flood had left only a sea of ice behind, or if the earth had the incorruptible hardness of jasper; let those who think the earth would be more beautiful after being changed into a crystal ball be transformed by Medusa's stare into a diamond statue!

But the objects chosen by the first physicists to explore the validity of a quantitative description—that is, the ideal pendulum with its conservative motion, simple machines, planetary orbits, etc.—were found to correspond to a *unique* mathematical description that actually reproduced the divine ideality of Aristotle's heavenly bodies.

Like Aristotle's gods, the objects of classical dynamics are concerned only with themselves. They can learn nothing from the outside. At any instant, each point in the system knows all it will ever need to know—that is, the distribution of masses in space and their velocities. Each state contains the whole truth concerning all possible other states, and each can be used to predict the others, whatever their respective positions on the time axis. In this sense, this description leads to a tautology, since both future and past are contained in the present.

The radical change in the outlook of modern science, the transition toward the temporal, the multiple, may be viewed as the reversal of the movement that brought Aristotle's heaven to earth. Now we are bringing earth to heaven. We are discovering the primacy of time and change, from the level of elementary particles to cosmological models.

Both at the macroscopic and microscopic levels, the natural sciences have thus rid themselves of a conception of objective reality that implied that novelty and diversity had to be denied in the name of immutable universal laws. They have rid themselves of a fascination with a rationality taken as closed and a knowledge seen as nearly achieved. They are now open to the unexpected, which they no longer define as the result of imperfect knowledge or insufficient control.

This opening up of science has been well defined by Serge Moscovici as the "Keplerian revolution," to distinguish it from the "Copernican revolution" in which the idea of an absolute point of view was maintained. In many of the passages cited in the Introduction to this book, science was likened to a "disenchantment" of the world. Let us quote Moscovici's description of the changes going on in the sciences today:

> Science has become involved in this adventure, our adventure, in order to renew everything it touches and warm all that it penetrates—the earth on which we live and the truths which enable us to live. At each turn it is not the echo of a demise, a bell tolling for a passing away that is heard, but the voice of rebirth and beginning, ever afresh, of mankind and materiality, fixed for an instant in their ephemeral permanence. That is why the great dis-

coveries are not revealed on a deathbed like that of Copernicus, but offered like Kepler's on the road of dreams and passion.[20]

The Creative Course of Time

It is often said that without Bach we would not have had the "St. Matthew Passion" but that relativity would have been discovered without Einstein. Science is supposed to take a deterministic course, in contrast with the unpredictability involved in the history of the arts. When we look back on the strange history of science, three centuries of which we have tried to outline, we may doubt the validity of such assertions. There are striking examples of facts that have been ignored because the cultural climate was not ready to incorporate them into a consistent scheme. The discovery of chemical clocks probably goes back to the nineteenth century, but their result seemed to contradict the idea of uniform decay to equilibrium. Meteorites were thrown out of the Vienna museum because there was no place for them in the description of the solar system. Our cultural environment plays an active role in the questions we ask, but beyond matters of style and social acceptance, we can identify a number of questions to which each generation returns.

The question of time is certainly one of those questions. Here we disagree somewhat with Thomas Kuhn's analysis of the formation of "normal" science.[21] Scientific activity best corresponds to Kuhn's view when it is considered in the context of the contemporary university, in which research and the training of future researchers is combined. Kuhn's analysis, if it is taken as a description of science in general, leading to conclusions about what knowledge must be, can be reduced to a new psychosocial version of the positivist conception of scientific development, namely, the tendency to increasing specialization and compartmentalization; the identification of "normal" scientific behavior with that of the "serious," "silent" researcher who wastes no time on "general" questions about the overall significance of his research but sticks to specialized problems; and the essential independence of scientific development from cultural, economic, and social problems.

The academic structure in which the "normal science" described by Kuhn came into being took shape in the nineteenth century. Kuhn emphasizes that it is by repeating in the form of exercises solutions to the paradigmatic problems of previous generations that students learn the concepts upon which research is based. It is in this way that they are given the criteria that define a problem as interesting and a solution as acceptable. The transition from student to researcher takes place gradually; the scientist continues to solve problems using similar techniques.

Even in our time, for which Kuhn's description has the greatest relevance, it refers to only one specific aspect of scientific activity. The importance of this aspect varies according to the individual researchers and the institutional context.

In Kuhn's view the transformation of a paradigm appears as a crisis: instead of remaining a silent, almost invisible rule, instead of remaining unspoken, the paradigm is actually questioned. Instead of working in unison, the members of the community begin to ask "basic" questions and challenge the legitimacy of their methods. The group, which by training was homogeneous, now diversifies. Different points of view, cultural experiences, and philosophic convictions are now expressed and often play a decisive role in the discovery of a new paradigm. The emergence of the new paradigm further increases the vehemence of the debate. The rival paradigms are put to the test until the academic world determines the victor. With the appearance of a new generation of scientists, silence and unanimity take over again. New textbooks are written, and once again things "go without saying."

In this view the driving force behind scientific innovation is the intensely conservative behavior of scientific communities, which stubbornly apply to nature the same techniques, the same concepts, and always end up by encountering an equally stubborn resistance from nature. When nature is eventually seen as refusing to express itself in the accepted language, the crisis explodes with the kind of violence that results from a breach of confidence. At this stage, all intellectual resources are concentrated on the search for a new language. Thus scientists have to deal with crises imposed upon them against their will.

The questions we have investigated have led us to emphasize

aspects that differ considerably from those to which Kuhn's description applies. We have dwelled on continuities, not the "obvious" continuities but the hidden ones, those involving difficult questions rejected by many as illegitimate or false but that keep coming back generation after generation—questions such as the dynamics of complex systems, the relation of the irreversible world of chemistry and biology with the reversible description provided by classical physics. In fact, the interest of such questions is hardly surprising. To us, the problem is rather to understand how they could ever have been neglected after the work of Diderot, Stahl, Venel, and others.

The past one hundred years have been marked by several crises that correspond closely to the description given by Kuhn—none of which were sought by scientists. Examples are the discovery of the instability of elementary particles, or of the evolving universe. However, the recent history of science is also characterized by a series of problems that are the consequences of deliberate and lucid questions asked by scientists who knew that the questions had both scientific and philosophical aspects. Thus scientists are not *doomed* to behave like "hypnons"!

It is important to point out that the new scientific development we have described, the incorporation of irreversibility into physics, is not to be seen as some kind of "revelation," the possession of which would set its possessor apart from the cultural world he lives in. On the contrary, this development clearly reflects both the internal logic of science and the cultural and social context of our time.

In particular, how can we consider as accidental that the rediscovery of time in physics is occurring at a time of extreme acceleration in human history? Cultural context cannot be the complete answer, but it cannot be denied either. We have to incorporate the complex relations between "internal" and "external" determinations of the production of scientific concepts.

In the preface of this book, we have emphasized that its French title *(La nouvelle alliance)* expresses the coming together of the "two cultures". Perhaps the confluence is nowhere as clear as in the problem of the microscopic foundations of irreversibility we have studied in Book Three.

As mentioned repeatedly, both classical and quantum mechanics are based on arbitrary initial conditions and deterministic laws (for trajectories or wave functions). In a sense, laws made simply explicit what was already present in the initial conditions. This is no longer the case when irreversibilty is taken into account. In this perspective, initial conditions arise from previous evolution and are transformed into states of the same class through subsequent evolution.

We come therefore close to the central problem of Western ontology: the relation between Being and Becoming. We have given a brief account of the problem in Chapter III. It is remarkable that two of the most influential works of the century were precisely devoted to this problem. We have in mind Whitehead's *Process and Reality* and Heidegger's *Sein und Zeit*. In both cases, the aim is to go beyond the identification of Being with timelessness, following the *Voie Royale* of western philosophy since Plato and Aristotle.[22]

But obviously, we cannot reduce Being to Time, and we cannot deal with a Being devoid of any temporal connotation. The direction which the microscopic theory of irreversibility takes gives a new content to the speculations of Whitehead and Heidegger.

It would go beyond the aim of this book to develop this problem in greater detail; we hope to do it elsewhere. Let us notice that initial conditions, as summarized in a state of the system, are associated with Being; in contrast, the laws involving temporal changes are associated with Becoming.

In our view, Being and Becoming are not to be opposed one to the other: they express two related aspects of reality.

A state with broken time symmetry arises from a law with broken time symmetry, which propagates it into a state belonging to the same category.

In a recent monograph *(From Being to Becoming),* one of the authors concluded in the following terms: "For most of the founders of classical science—even for Einstein—science was an attempt to go beyond the world of appearances, to reach a timeless world of supreme rationality—the world of Spinoza. But perhaps there is a more subtle form of reality that involves both laws and games, time and eternity."

This is precisely the direction which the microscopic theory of irreversible processes is taking.

The Human Condition

We agree completely with Herman Weyl:

> Scientists would be wrong to ignore the fact that theoretical construction is not the only approach to the phenomena of life; another way, that of understanding from within (interpretation), is open to us. . . . Of myself, of my own acts of perception, thought, volition, feeling and doing, I have a direct knowledge entirely different from the theoretical knowledge that represents the "parallel" cerebral processes in symbols. This inner awareness of myself is the basis for the understanding of my fellowmen whom I meet and acknowledge as beings of my own kind, with whom I communicate sometimes so intimately as to share joy and sorrow with them.[23]

Until recently, however, there was a striking contrast. The external universe appeared to be an automaton following deterministic causal laws, in contrast with the spontaneous activity and irreversibility we experience. The two worlds are now drawing closer together. Is this a loss for the natural sciences?

Classical science aimed at a "transparent" view of the physical universe. In each case you would be able to identify a cause and an effect. Whenever a stochastic description becomes necessary, this is no longer so. We can no longer speak of causality in each individual experiment; we can only speak about statistical causality. This has, in fact, been the case ever since the advent of quantum mechanics, but it has been greatly amplified by recent developments in which randomness and probability play an essential role, even in classical dynamics or chemistry. Therefore, the modern trend as compared to the classical one leads to a kind of "opacity" as compared to the transparency of classical thought.

Is this a defeat for the human mind? This is a difficult question. As scientists, we have no choice; we cannot describe for you the world as we would like to see it, but only as we are able to see it through the combined impact of experimental results and new theoretical concepts. Also, we believe that this new situation reflects the situation we seem to find in our own mental activity. Classical psychology centered around conscious, transparent activity; modern psychology attaches much weight to the opaque functioning of the unconscious. Perhaps this is an image of the basic features of human existence. Remember Oedipus, the lucidity of his mind in front of the sphinx and its opacity and darkness when confronted with his own origins. Perhaps the coming together of our insights about the world around us and the world inside us is a satisfying feature of the recent evolution in science that we have tried to describe.

It is hard to avoid the impression that the distinction between what exists in time, what is irreversible, and, on the other hand, what is outside of time, what is eternal, is at the origin of human symbolic activity. Perhaps this is especially so in artistic activity. Indeed, one aspect of the transformation of a natural object, a stone, to an object of art is closely related to our impact on matter. Artistic activity breaks the temporal symmetry of the object. It leaves a mark that translates our temporal dissymmetry into the temporal dissymmetry of the object. Out of the reversible, nearly cyclic noise level in which we live arises music that is both stochastic and time-oriented.

The Renewal of Nature

It is quite remarkable that we are at a moment both of profound change in the scientific concept of nature and of the structure of human society as a result of the demographic explosion. As a result, there is a need for new relations between man and nature and between man and man. We can no longer accept the old a priori distinction between scientific and ethical values. This was possible at a time when the external world and our internal world appeared to conflict, to be nearly orthogonal. Today we know that time is a construction and therefore carries an ethical responsibility.

The ideas to which we have devoted much space in this book—the ideas of instability, of fluctuation—diffuse into the social sciences. We know now that societies are immensely complex systems involving a potentially enormous number of bifurcations exemplified by the variety of cultures that have evolved in the relatively short span of human history. We know that such systems are highly sensitive to fluctuations. This leads both to hope and a threat: hope, since even small fluctuations may grow and change the overall structure. As a result, individual activity is not doomed to insignificance. On the other hand, this is also a threat, since in our universe the security of stable, permanent rules seems gone forever. We are living in a dangerous and uncertain world that inspires no blind confidence, but perhaps only the same feeling of qualified hope that some Talmudic texts appear to have attributed to the God of Genesis:

> Twenty-six attempts preceded the present genesis, all of which were destined to fail. The world of man has arisen out of the chaotic heart of the preceding debris; he too is exposed to the risk of failure, and the return to nothing. "Let's hope it works" [*Halway Sheyaamod*] exclaimed God as he created the World, and this hope, which has accompanied all the subsequent history of the world and mankind, has emphasized right from the outset that this history is branded with the mark of radical uncertainty.[24]

NOTES

Introduction

1. I. BERLIN, *Against the Current,* selected writings ed. H. Hardy (New York: The Viking Press, 1980), p. xxvi.
2. See TITUS LUCRETIUS CARUS, *De Natura Rerum,* Book I, v. 267–70. ed. and comm. C. Bailey (Oxford: Oxford University Press 1947, 3 vols.)
3. R. LENOBLE, *Histoire de l'idée de nature* (Paris: Albin Michel, 1969).
4. B. PASCAL, "Pensées," frag. 792, in *Oeuvres Complètes* (Paris: Brunschwig-Boutroux-Gazier, 1904–14).
5. J. MONOD, *Chance and Necessity* (New York: Vintage Books, 1972), pp. 172–73.
6. G. VICO, *The New Science,* trans. T. G. Bergin and M. H. Fisch (New York: 1968), par. 331.
7. J. P. VERNANT et al., *Divination et rationalité,* esp. J. BOTTERO, "Symptômes, signes, écritures" (Paris: Seuil, 1974).
8. A. KOYRÉ, *Galileo Studies* (Hassocks, Eng.: The Harvester Press, 1978).
9. K. POPPER, *Objective Knowledge* (Oxford: Clarendon Press, 1972).
10. P. FORMAN, "Weimar Culture, Causality and Quantum Theory, 1918–1927; Adaptation by German Physicists and Mathematicians to an Hostile Intellectual Environment," *Historical Studies in Physical Sciences,* Vol. 3 (1971), pp. 1–115.
11. J. NEEDHAM and C. A. RONAN, *A Shorter Science and Civilization in China,* Vol. I (Cambridge: Cambridge University Press, 1978), p. 170.
12. A. EDDINGTON, *The Nature of the Physical World* (Ann Arbor: University of Michigan Press, 1958), pp. 68–80.
13. Ibid., p. 103.
14. BERLIN, op. cit., p. 109.
15. K. POPPER, *Unended Quest* (La Salle, Ill.: Open Court Publishing Company, 1976), pp. 161–62.
16. G. BRUNO, 5th dialogue, "De la causa," *Opere Italiane,* I (Bari: 1907); cf. I. LECLERC, *The Nature of Physical Existence* (London: George Allen & Unwin, 1972).

17. P. VALÊRY, *Cahiers,* (2 vols.) ed. Mrs. Robinson-Valêry, (Paris: Gallimard, 1973–74).
18. E. SCHRÖDINGER, "Are there Quantum Jumps?" *The British Journal for the Philosophy of Science,* Vol. III (1952), pp. 109–10; this text has been quoted with indignation by P. W. Bridgmann in his contribution to *Determinism and Freedom in the Age of Modern Science,* ed. S. Hook (New York: New York University Press, 1958).
19. A. EINSTEIN, "Prinzipien der Forschung, Rede zur 60. Geburtstag von Max Planck" (1918) in *Mein Weltbild,* Ullstein Verlag 1977, pp. 107–10, trans. *Ideas and Opinions* (New York: Crown, 1954), pp. 224–27.
20. F. DÜRRENMATT, *The Physicists.* (New York: Grove, 1964).
21. S. MOSCOVICI, *Essai sur l'histoire humaine de la nature,* Collection Champs (Paris: Flammarion, 1977).
22. Quoted in Needham and Ronan, op. cit., p. 87.
23. MONOD, op. cit., p. 180.

Chapter 1

1. J. T. DESAGULIERS, "The Newtonian System of the World, The Best Model of Government: an Allegorical Poem," 1728, quoted in H. N. FAIRCHILD, *Religious Trends in English Poetry,* Vol. I (New York: Columbia University Press, 1939), p. 357.
2. Ibid., p. 358.
3. Gerd Buchdahl emphasized and illustrated this ambiguity of the cultural influence of the Newtonian model in its dimensions both empirical *(Opticks)* and systematic *(Principia)* in *The Image of Newton and Locke in the Age of Reason,* Newman History and Philosophy of Science Series (London: Sheed & Ward, 1961).
4. *La Science et la diversité des cultures,* (Paris: UNESCO, PUF, 1974), pp. 15–16.
5. C. C. GILLISPIE, *The Edge of Objectivity* (Princeton, N.J.: Princeton University Press, 1970), pp. 199–200.
6. M. HEIDEGGER, *The Question Concerning Technology* (New York: Harper & Row, 1977), p. 20.
7. Ibid., p. 21.
8. Ibid., p. 16.
9. "The Coming of the Golden Age," *Paradoxes of Progress* (San Francisco: Freeman & Company, 1978).
10. See, for instance, P. DAVIES, *Other Worlds* (Toronto: J. M. Dent & Sons, 1980).

11. A. KOESTLER, *The Roots of Coincidence* (London: Hutchinson, 1972), pp. 138–39.
12. A. KOYRÉ, *Newtonian Studies* (Chicago: University of Chicago Press, 1968), pp. 23–24.
13. In "Race and History" (*Structural Anthropology II,* New York: Basic Books, 1976), Claude Lévi-Strauss discusses the conditions that lead to the Neolithic and Industrial revolutions. The model he introduces, involving chain reactions and catalysis (a process with kinetics characterized by threshold and amplification phenomena) attests to an affinity between the problems of stability and fluctuation we discuss in Chapter VI as well as certain themes of the "structural approach" in anthropology.
14. "Inside each society, the order of myth excludes dialogue: the group's myths are not discussed, they are transformed when they are thought to be repeated." C. LÉVI-STRAUSS, *L'Homme Nu* (Paris: Plon, 1971), p. 585. Thus mythical discourse is to be distinguished from critical (scientific and philosophic) dialogue more because of the practical conditions of its reproduction than because of an intrinsic inability of such or such emitter to think in a rational way. The practice of critical dialogue has given to the cosmological discourse claiming truthfulness its spectacular evolutive acceleration.
15. This is, of course, one of the main themes of Alexandre Koyré.
16. The definition of such an "absurdity" opposes the age-long idea that a sufficiently tricky device would permit one to cheat nature. See the efforts devoted by engineers till the twentieth century to the construction of perpetual-motion machines in A. Ord Hume, *Perpetual Motion: The History of an Obsession* (New York: St. Martin's Press, 1977).
17. Popper translated into a norm this excitement born out of the risks involved in the experimental games. He affirms, in *The Logic of Scientific Discovery,* that the scientific *must* look for the most "improbable" hypothesis—that is, the most risky one—to try to refute it as well as the corresponding theories.
18. R. FEYNMAN, *The Character of Physical Law* (Cambridge, Mass.: M.I.T. Press, 1967), second chapter.
19. J. NEEDHAM, "Science and Society in East and West," *The Grand Titration* (London: Allen & Unwin, 1969).
20. A. N. WHITEHEAD, *Science and the Modern World* (New York: The Free Press, 1967), p. 12.
21. NEEDHAM, op. cit., p. 308.
22. NEEDHAM, op. cit., p. 330.
23. R. HOOYKAAS emphasized this "dedivinization" of the world by the Christian metaphor of the world machine in *Religion and the*

Rise of Modern Science (Edinburgh and London: Scottish Academic Press, 1972), esp. pp. 14–16.

24. WHITEHEAD, op. cit., p. 54.
25. The famous text about nature being written in mathematical signs is to be found in *Il Saggiatore*. See also *The Dialogue Concerning the Two Chief World Systems,* 2nd rev. ed. (Berkeley: University of California Press, 1967).
26. At least it was triumphant in the academies created in France, Prussia, and Russia by absolute sovereigns. In *The Scientist's Role in Society* (Englewood Cliffs, N.J.: Foundations of Modern Sociology Series, Prentice-Hall, 1971), Ben David emphasized the distinction between physicists of these countries, dedicated to physics as a glamorous and purely theoretical science, and the English physicists immerged in a wealth of empirical and technical problems. Ben David proposed a connection between the fascination for a theoretical science and the relegation far from political power of the social class supporting the "scientific movement."
27. In his biography of d'Alembert—*Jean d'Alembert, Science and Enlightenment* (Oxford: Clarendon Press, 1970)—Thomas Hankins emphasized how closed and small was the first true scientific community, in the modern sense of the term, namely, that of the eighteenth-century physicists and mathematicians, and how intimate were their relations with the absolute sovereigns.
28. EINSTEIN, op. cit., pp. 225–26.
29. E. MACH, "The Economical Nature of Physical Inquiry," *Popular Scientific Lectures* (Chicago: Open Court Publishing Company, 1895), pp. 197–98.
30. J. DONNE, *An Anatomy of the World* wherein . . . the frailty and the decay of the whole world is represented (London, catalog of the British Museum, 1611).

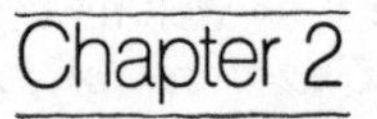

Chapter 2

1. On this point, see T. HANKINS, "The Reception of Newton's Second Law of Motion in the Eighteenth Century," *Archives Internationales d'Histoire des Sciences,* Vol. XX (1967), pp. 42–65, and I. B. COHEN, "Newton's Second Law and the Concept of Force in the Principia," *The Annus Mirabilis of Sir Isaac Newton, Tricentennial Celebration, The Texas Quarterly,* Vol. X, No. 3 (1967), pp. 25–157. The four following paragraphs rest, for what concerns atomism and the conservation theories,

on W. SCOTT, *The Conflict Between Atomism and Conservation Theory* (London: Macdonald, 1970).

2. A. KOYRÉ, *Galileo Studies* (Hassocks, Eng.: The Harvester Press, 1978), pp. 89–94.
3. In his history of mechanics—*The Science of Mechanics: A Critical and Historical Account of Its Development* (La Salle, Ill.: Open Court Publishing Company, 1960)—Ernst Mach laid stress on this dual filiation of modern dynamics of both the trajectories science and the engineer's computations.
4. This at least is the conclusion of today's historians who began the study of the impressive mass of Newton's "Alchemical Papers," which till now were ignored or disdained as "nonscientific." See B. J. DOBBS, *The Foundations of Newton's Alchemy* (Cambridge: Cambridge University Press, 1975); R. WESTFALL, "Newton and the Hermetic Tradition" in *Science, Medicine and Society,* ed. A. G. DEBUS (London: Heinemann, 1972); and R. WESTFALL, "The Role of Alchemy in Newton's Career," *Reason, Experiment and Mysticism,* ed. M. L. RIGHINI BONELLI and W. R. SHEA (London: Macmillan, 1975). As Lord Keynes, who played a crucial part in the collection of these papers, summarized (quoted in DOBBS, op. cit., p. 13), "Newton was not the first of the age of reason. He was the last of the Babylonians and Sumerians, the last great mind which looked out on the visible and intellectual world with the same eyes as those who began to build our intellectual inheritance rather less than 10,000 years ago."
5. DOBBS, op. cit., also examined the role of the "mediator" by which two substances are made "sociable." We may recall here the importance of the mediator in Goethe's *Elective Affinities* (Engl. trans. Greenwood 1976). For what concerns chemistry, Goethe was not far from Newton.
6. The story of Newton's "mistake" is told in HANKINS's, *Jean d'Alembert,* pp. 29–35.
7. G. L. BUFFON, "Réflexions sur la loi d'attraction," appendix to *Introduction à l'histoire des minéraux* (1774), Tome IX of *Oeuvres Complètes* (Paris: Garnier Frères), pp. 75, 77.
8. G. L. BUFFON, *Histoire naturelle. De la Nature, Seconde Vue* (1765), quoted in H. METZGER, *Newton, Stahl, Boerhaave et la doctrine chimique* (Paris: Blanchard, 1974), pp. 57–58.
9. A. THACKRAY describes the way French chemistry became Buffonian in *Atom and Power: An Essay on Newtonian Matter Theory and the Development of Chemistry* (Cambridge, Mass.: Harvard University Press, 1970), pp. 199–233. Berthollet's *Statique chimique* accomplished Buffon's program and also

closed it, since his disciples gave up the attempt to understand chemical reactions in terms compatible with Newtonian concepts.

10. We do not wish to try to explain here the reasons of Newton's triumph in France, nor of its fall, but to emphasize the at least chronological connection between these events and the stages of the process of professionalization of science. See M. CROSLAND, *The Society of Arcueil: A View of French Science at the Time of Napoléon* (London: Heinemann, 1960), as well as his *Gay Lussac* (Cambridge: Cambridge University Press, 1978).
11. Thomas Kuhn made of this role of scientific institutions, taking over the formation of the future scientists—that is assuring their own reproduction, the main characteristic of scientific activity as we know it today. This problem has also been approached by M. Crosland, R. Hahn, and W. Farrar in *The Emergence of Science in Western Europe,* ed. M. CROSLAND (London: Macmillan, 1975).
12. The role of "mundane" interest so despised by philosophers such as Gaston Bachelard in France should be taken as the sign of the open character of eighteenth-century science. In a way, we can truly speak about a regression during the nineteenth century, at least for what concerns the scientific *culture*. And we could learn today from the multiplicity of local academies and circles where scientific matters were discussed by nonprofessionals.
13. Quoted in J. SCHLANGER, *Les métaphores de l'organisme* (Paris: Vrin, 1971), p. 108.
14. J. C. MAXWELL, *Science and Free Will,* in CAMPBELL and GARNETT, op. cit., p. 443. L. CAMPBELL & W. GARNETT, *The Life of James Clerk Maxwell* (London, Macmillan, 1882).
15. This problem is one of the main themes of French philosopher Michel Serres. See, for instance, "Conditions" in his *La naissance de la physique dans le texte de Lucrèce* (Paris: Minuit, 1977). Some texts by M. Serres are now available in English translation, thanks to the pious zeal of the French Studies Department of Johns Hopkins University. See M. SERRES, *Hermes: Literature, Science, Philosophy.* (Baltimore: The Johns Hopkins University Press, 1982.)
16. See, about the fate of Laplace's demon, E. CASSIRER, *Determinism and Interdeterminism in Modern Physics* (New Haven, Conn.: Yale University Press, 1956), pp. 3–25.

Chapter 3

1. R. NISBET, *History of the Idea of Progress* (New York: Basic Books, 1980), p. 4.
2. D. DIDEROT, *d'Alembert's Dream* (Harmondsworth:, Eng.: Penguin Books, 1976), pp. 166–67.
3. D. DIDEROT, *"Conversation Between d'Alembert and Diderot," d'Alembert's Dream,* pp. 158–59.
4. D. DIDEROT, *Pensées sur l'Interprétation de la Nature* (1754), *Oeuvres Complètes,* Tome II (Paris: Garnier Frères, 1875), p. 11.
5. Diderot ascribes this opinion to the physician Bordeu in the *Dream.*
6. See, for instance, A. LOVEJOY, *The Great Chain of Beings* (Cambridge, Mass.: Harvard University Press, 1973).
7. The historian Gillispie proposed a relation between the protest against mathematical physics, as popularized by Diderot in the *Encyclopédie,* and the revolutionaries' hostility against this official science, as manifested by the closure of the Academy and Lavoisier's death. This is a very controversial point, but what is sure is that the Newtonian triumph in France coincides with the Napoleonic institutions, spelling the final victory of state academy over craftsmen (see C. C. GILLISPIE, "The Encyclopedia and the Jacobin Philosophy of Science: A Study in Ideas and Consequences," *Critical Problems in the History of Science,* ed. M. CLAGETT (Madison, Wis.: University of Wisconsin Press, 1959), pp. 255–89.
8. G. E. STAHL, "Véritable Distinction à établir entre le mixte et le vivant du corps humain," *Oeuvres médicophilosophiques et pratiques,* Tome II (Montpellier: Pitrat et Fils, 1861), esp. pp. 279–82.
9. See J. SCHLANGER, *Les métaphores de l'organisme,* for a description of the transformation of the meaning of "organization" between Stahl and the Romanticists.
10. *Philosophy of Nature,* §261.
11. This is Knight's conclusion in "The German Science in the Romantic Period," *The Emergence of Science in Western Europe.*
12. H. BERGSON, *La pensée et le mouvant* in *Oeuvres* (Paris: Éditions du Centenaire, PUF, 1970), p. 1285; trans. *The Creative Mind* (Totowa, N.J.: Littlefield, Adams, 1975), p. 42.
13. Ibid., p. 1287; trans., p. 44.
14. Ibid., p. 1286; trans., p. 44.

15. H. BERGSON, *L'évolution créatrice* in *Oeuvres,* p. 784; trans. *Creative Evolution* (London: Macmillan, 1911), p. 361.
16. Ibid., p. 538; trans., p. 54.
17. Ibid., p. 784; trans., p. 361.
18. BERGSON, *La pensée et le mouvant,* p. 1273; trans., p. 32.
19. Ibid., p. 1274; trans., p. 33.
20. A. N. WHITEHEAD, *Science and the Modern World,* p. 55.
21. A. N. WHITEHEAD, *Process and Reality: An Essay in Cosmology* (New York: The Free Press, 1969), p. 20.
22. Ibid., p. 26.
23. Joseph Needham and C. H. Waddington both acknowledged the importance of Whitehead's influence for what concerns their endeavor to describe in a *positive* way the organism as a whole.
24. H. HELMHOLTZ, *Uber die Erhaltung der Kraft* (1847), trans. in S. BRUSH, *Kinetic Theory,* Vol. I, *The Nature of Gases and Heat* (Oxford: Pergamon Press, 1965), p. 92. See also Y. ELKANA, *The Discovery of the Conservation of Energy* (London: Hutchinson Educational, 1974) and P. M. HEIMANN, "Helmholtz and Kant: The Metaphysical Foundations of Über die Erhaltung der Kraft," *Studies in the History and Philosophy of Sciences,* Vol. 5 (1974), pp. 205–38.
25. H. REICHENBACH, *The Direction of Time* (Berkeley: University of California Press, 1956), pp. 16–17.

Chapter 4

1. W. SCOTT, About the novelty of these problems, see *The Conflict Between Atomism and Conservation Theory,* Book II, and about the industrial context where these concepts were created, D. CARDWELL, *From Watt to Clausius* (London, Heinemann, 1971). Particularly interesting in this respect is the convergence between on one hand the need determined by industrial problems and on the other the positivist simplifications by operational definitions.
2. J. HERIVEL, *Joseph Fourier: The Man and the Physicist* (Oxford: Clarendon Press, 1975). In this biography we learn the following curious information: Fourier would have brought back from his trip with Bonaparte to Egypt a sickness causing permanent deperditions of heat.
3. See, more particularly, the introduction to Comte's *Philosophie Première* (Paris: Herman, 1975), "Auguste Comte auto-traduit dans l'encyclopédie" in *La Traduction* (Paris: Minuit, 1974) and "Nuage," *La Distribution* (Paris: Minuit, 1977).

4. C. SMITH, "Natural Philosophy and Thermodynamics: William Thomson and the Dynamical Theory of Heat," *The British Journal for the Philosophy of Science,* Vol. 9 (1976), pp. 293–319 and M. CROSLAND and C. SMITH, "The Transmission of Physics from France to Britain, 1800–1840," *Historical Studies in the Physical Sciences,* Vol. 9 (1978), pp. 1–61.
5. For what follows, see Y. ELKANA, *The Discovery of the Conservation of Energy Principle,* as well as the famous paper by Thomas Kuhn, "Energy conservation as an Example of Simultaneous Discovery," originally published in *Critical Problems in the History of Science* and recently in T. KUHN, *The Essential Tension* (Chicago: University of Chicago Press, 1977).
6. ELKANA followed the slow crystallization of the concept of energy; see his book and "Helmholtz's Kraft: An Illustration of Concepts in Flux," *Historical Studies in the Physical Sciences,* Vol. 2 (1970), pp. 263–98.
7. J. JOULE, "Matter, Living Force and Heat," *The Scientific Papers of James Prescott Joule,* Vol. 1 (London: Taylor & Francis, 1884), pp. 265–76 (quotation, p. 273).
8. The English translations of Mayer's two great papers, "On the Forces of Inorganic Nature" and "The Motions of Organisms and Their Relation to Metabolism," are in *Energy: Historical Development of a Concept,* ed. R. B. LINDSAY (Stroudsburg, Pa.: Benchmarks Papers on Energy 1, Dowden, Hutchinson & Ross, 1975).
9. E. BENTON, "Vitalism in the Nineteenth Century Scientific Thought: A Typology and Reassessment," *Studies in History and Philosophy of Science,* Vol. 5 (1974), pp. 17–48.
10. H.HELMHOLTZ, "Über die Erhaltung der Kraft," op. cit., pp. 90–91.
11. G. DELEUZE, *Nietzsche et la philosophie* (Paris: PUF, 1973), pp. 48–55.
12. In this study of Zola's "Docteur Pascal," *Feux et signaux de brume* Paris: Grasset (1975), p. 109, Michel Serres wrote: "The century that was practically drawing to a close when the novel appeared had opened with the majestic stability of the solar system, and was now filled with dismay at the relentless degradations of fire. Hence the fierce, positive dilemma: perfect cycle without residue, eternal and positively valued, i.e., the cosmology of the sun; or else a missed cycle, losing its difference, irreversible, historical and despised—a cosmology, a thermogony of fire which must either be extinguished or destroyed, without alternative. One dreams of Laplace, whilst Carnot and the others have forever smashed the cubby-hole, the niche, where one

could sleep in peace; one is dreaming, that is certain: then cultural archaisms having returned through another door, through another opening of the same door, are powerfully reawakened: immortal flame, purifying blaze or evil fire?"

13. The continuity between Carnot father and son has been emphasized by Cardwell (*From Watt to Clausius*) and Scott (*The Conflict Between Atomism and Conservation Theory*).
14. P. DAVIES, *The Runaway Universe* (New York: Penguin Books, 1980), p. 197.
15. F. DYSON, "Energy in the Universe," *Scientific American*, Vol. 225 (1971), pp. 50–59.
16. What was particularly important was to grasp that, unlike what happens in mechanics, it is not just any situation of a thermodynamic system that can be characterized as a "state"; quite the contrary. See E. DAUB, "Entropy and Dissipation," *Historical Studies in the Physical Sciences*, Vol. 2 (1970), pp. 321–54.
17. In his autobiography, *Scientific Autobiography* (London: Williams & Norgate, 1950), Max Planck recalled how isolated he had been when he first emphasized the peculiarity of heat and pointed out that it is the conversion of heat into another form of energy that raises the irreversibility problem. Energeticists such as Ostwald wanted all forms of energy to be given the same status. For them, the fall of a body between two altitude levels takes place by virtue of the same kind of productive difference as the passage of heat between two bodies at different temperatures. Thus, Ostwald's comparison did away with the crucial distinction between an ideally reversible process, such as the mechanical motion, and an intrinsically irreversible one, such as heat diffusion. By doing so, he was actually taking up a position similar to what we have attributed to Lagrange: where Lagrange considered conservation of energy as a property belonging only to ideal cases but also the only one capable of being treated rigorously, Ostwald held conservation of energy as the property of any natural transformation, but defined conservation of energy *differences* (required by all transformation since only a difference can produce another difference) as an abstract ideal, but the sole object for a rational science.
18. The splitting of the entropy variation into two different terms was introduced in 1. Prigogine, *Etude thermodynamique des Phénomènes irréversibles*, Thèse d'agrégation présentée à la faculté des sciences de l'Université Libre de Bruxelles 1945 (Paris: Dunod, 1947).
19. R. CLAUSIUS, *Ann. Phys.*, Vol. 125 (1865), p. 353.
20. M. PLANCK, "The Unity of the Physical Universe," *A Survey of*

Physics, Collection of Lectures and Essays (New York: E. P. Dutton, 1925), p. 16.

21. R. CAILLOIS, "La dissymétrie," *Cohérences aventureuses,* Collection Idées (Paris: Gallimard, 1973), p. 198.

Chapter 5

1. For what concerns the content of this and the following chapter, see P. GLANSDORFF and I. PRIGOGINE, *Thermodynamic Theory of Structure, Stability and Fluctuations* (New York: John Wiley & Sons, 1971) and G. NICOLIS and I. PRIGOGINE, *Self-Organization in Non-Equilibrium Systems* (New York: John Wiley & Sons, 1977), where further references may be found.
2. F. NIETZSCHE, *Der Wille zur Macht, Sämtliche Werke* (Stuttgart: Kroner, 1964), aphorism 630.
3. Which precise content can be given to the general law of entropy growth? For a theoretician physicist such as de Donder, chemical activity, which appeared obscure and inaccessible to the rational approach of mechanics, was mysterious enough to become the synonym of the irreversible process. Thus chemistry, whose question physicists had never truly answered, and the new enigma of irreversibility came to join in a challenge not to be ignored anymore. See Th. De Donder, *L'Affinité* (Paris: Gauthier-Villars, 1962) and L. Onsager *Phys. Rev. 37,* 405 (1931).
4. M. SERRES, *La naissance de la physique dans le texte de Lucrèce, op. cit.*
5. For more details concerning chemical oscillations, see A. WINFREE, "Rotating Chemical Reactions," *Scientific American,* Vol. 230 (1974), pp. 82–95.
6. A. GOLDBETER and G. NICOLIS, "An Allosteric Model with Positive Feedback Applied to Glycolytic Oscillations," *Progress in Theoretical Biology,* Vol. 4 (1976), pp. 65–160; A. GOLDBETER and S. R. CAPLAN, "Oscillatory Enzymes," *Annual Review of Biophysics and Bioengineering,* Vol. 5 (1976), pp. 449–73..
7. B. HESS, A. BOITEUX, and J. KRÜGER, "Cooperation of Glycolytic Enzymes," *Advances in Enzyme Regulation,* Vol. 7 (1969), pp. 149–67; see also B. HESS, A. GOLDBETER, and R. LEFEVER, "Temporal, Spatial and Functional Order in Regulated Biochemical Cellular Systems," *Advances in Chemical Physics,* Vol. XXXVIII (1978), pp. 363–413.
8. B. HESS, *Ciba Foundation Symposium,* Vol. 31 (1975), p. 369.

9A. G. GERESCH, "Cell Aggregation and Differentiation in *Dic-*

tyostelium Discoideum," in *Developmental Biology,* Vol. 3 (1968), pp. 157–197.

9B. A. GOLDBETER and L. A. SEGEL, "Unified Mechanism for Relay and Oscillation of Cyclic AMP in *Dictyostelium Discoideum,*" *Proceedings of the National Academy of Sciences,* Vol. 74 (1977), pp. 1543–47.

10. See M. GARDNER, *The Ambidextrous Universe* (New York: Charles Scribner's Sons, 1979).

11. D.K. KONDEPUDI and I. PRIGOGINE, *Physica,* Vol. 107A (1981), pp. 1–24; D. K. KONDEPUDI, *Physica,* Vol. 115A (1982), pp. 552–66. It could even be that chemistry may bring to the macroscopic scale the violation of parity in weak forces; D. K. KONDEPUDI and G. W. NELSON, *Physical Review Letters,* Vol. 50, No. 14 (1983), pp. 1023–26.

12. R. LEFEVER and W. HORSTHEMKE, "Multiple Transitions Induced by Light Intensity Fluctuations in Illuminated Chemical Systems," *Proceedings of the National Academy of Sciences,* Vol. 76 (1979), pp. 2490–94. See also W. HORSTHEMKE and M. MALEK MANSOUR, "Influence of External Noise on Nonequilibrium Phase Transitions," *Zeitschrift für Physik B,* Vol. 24 (1976), pp. 307–13; L. ARNOLD, W. HORSTHEMKE, and R. LEFEVER, "White and Coloured External Noise and Transition Phenomena in Nonlinear Systems," *Zeitschrift für Physik B,* Vol. 29 (1978), pp. 367–73; W. HORSTHEMKE, "Nonequilibrium Transitions Induced by External White and Coloured Noise," *Dynamics of Synergetic Systems,* ed. H. HAKEN (Berlin: Springer Verlag, 1980); for an application to a biological problem, R. LEFEVER and W. HORSTHEMKE, "Bistability in Fluctuating Environments: Implication in Tumor Immunology," *Bulletin of Mathematic Biology,* Vol. 41 (1979).

13. H. L. SWINNEY and J. P. GOLLUB, "The Transition to Turbulence," *Physics Today,* Vol. 31, No. 8 (1978), pp. 41–49.

14. M. J. FEIGENBAUM, "Universal Behavior in Nonlinear Systems," *Los Alamos Science,* No 1 (Summer 1980), pp. 4–27.

15. The concept of chreod is part of the qualitative description of embryological development Waddington proposed more than twenty years ago. It is truly a bifurcating evolution: a progressive exploration along which the embryo grows in an "epigenetic landscape" where coexist stable segments and segments where a choice among several developmental paths is possible. See C. H. WADDINGTON, *The Strategy of the Genes* (London: Allen & Unwin, 1957). C. H. Waddington's chreods are also a central reference in René Thom's biological thought. They could thus become a meeting point for two approaches: the one we are presenting, starting from local mechanisms and exploring the

spectrum of collective behaviors they can generate; and Thom's, starting from global mathematical entities and connecting the qualitatively distinct forms and transformations they imply with the phenomenological description of morphogenesis.

16. S. A. KAUFFMAN, R. M. SHYMKO, and K. TRABERT, "Control of Sequential Compartment Formation in Drosophila," *Science,* Vol. 199 (1978), pp. 259–69.
17. H. BERGSON, *Creative Evolution,* pp. 94–95.
18. See C. H. WADDINGTON, *The Evolution of an Evolutionist* (Edinburgh: Edinburgh University Press, 1975) and P. WEISS, "The Living System: Determinism Stratified," *Beyond Reductionism,* ed. A. KOESTLER and J. R. SMYTHIES (London: Hutchinson, 1969).
19. D. E. KOSHLAND, "A Model Regulatory System: Bacterial Chemotaxis," *Physiological Review,* Vol. 59, No. 4, pp. 811–62.

1. G. NICOLIS and I. PRIGOGINE, *Self-Organization in Nonequilibrium Systems* (New York: John Wiley & Sons, 1977).
2. F. BARAS, G. NICOLIS, and M. MALEK MANSOUR, "Stochastic Theory of Adiabatic Explosion," *Journal of Statistical Physics,* Vol. 32, No. 1 (1983), pp. 1.
3. See, for example, M. MALEK MANSOUR, C. VAN DEN BROECK, G. NICOLIS, and J. W. TURNER, *Annals of Physics,* Vol. 131, No. 2 (1981), p. 283.
4. J. L. DENEUBOURG, "Application de l'ordre par fluctuation à la description de certaines étapes de la construction du nid chez les termites," *Insectes Sociaux, Journal International pour l'étude des Arthropodes sociaux,* Tome 24, No. 2 (1977), pp. 117–30. This first model is presently being extended in connection with new experimental studies; O. H. BRUINSMA, "An Analysis of Building Behaviour of the Termite *macrotermes subhyalinus,*" *Proceedings of the VIII Congress IUSSI* (Waegeningen, 1977).
5. R. P. GARAY and R. LEFEVER, "A Kinetic Approach to the Immunology of Cancer: Stationary States Properties of Effector-Target Cell Reactions," *Journal of Theoretical Biology,* Vol. 73 (1978), pp. 417–38, and private communication.
6. P. M. ALLEN, "Darwinian Evolution and a Predator-Prey Ecology," *Bulletin of Mathematical Biology,* Vol. 37 (1975), pp. 389–405; "Evolution, Population and Stability," *Proceedings of the National Academy of Sciences,* Vol. 73, No. 3 (1976), pp. 665–68. See also R. CZAPLEWSKI, "A Methodology for Eval-

uation of Parent-Mutant Competition," *Journal for Theoretical Biology,* Vol. 40 (1973), pp. 429–39.

7. See, for the present state of this work, M. EIGEN and P. SCHUSTER, *The Hypercycle* (Berlin: Springer Verlag, 1979).
8. R. MAY in *Science,* Vol. 186 (1974), pp. 645–47; see also R. MAY, "Simple Mathematical Models with very Complicated Dynamics," *Nature,* Vol. 261 (1976), pp. 459–67.
9. M. P. HASSELL, *The Dynamics in Arthropod Predator-Prey Systems* (Princeton, N.J.: Princeton University Press, 1978).
10. B. HEINRICH, "Artful Diners," *Natural History,* Vol. 89, No. 6 (1980), pp. 42–51, esp. quote, p. 42.
11. M. LOVE, "The Alien Strategy," *Natural History,* Vol. 89, No. 5 (1980), pp. 30–32.
12. J. L. DENEUBOURG and P. M. ALLEN, "Modèles théoriques de la division du travail des les sociétés d'insectes," *Académie Royale de Belgique, Bulletin de la Classe des Sciences,* Tome LXII (1976), pp. 416–29; P. M. ALLEN, "Evolution in an Ecosystem with Limited Resources," op. cit., pp. 408–15.
13. E. W. MONTROLL, "Social Dynamics and the Quantifying of Social Forces," *Proceedings of the National Academy of Sciences,* Vol. 75, No. 10 (1978), pp. 4633–37.
14. P. M. ALLEN and M. SANGLIER, "Dynamic Model of Urban Growth," *Journal for Social and Biological Structures,* Vol. 1 (1978), pp. 265–80, and "Urban Evolution, Self-Organization and Decision-making," *Environment and Planning A,* Vol. 13 (1981), pp. 167–83.
15. C. H. WADDINGTON, *Tools for Thought,* (St. Albans, Eng.: Paladin, 1976), p. 228.
16. S. J. GOULD, *Ontogeny and Phylogeny,* op. cit. Belknap Press Harvard University Press, 1977.
17. C. LÉVI-STRAUSS, "Methodes et enseignement," *Anthropologie structurale* (Paris: Plon), pp. 311–17.
18. See, for instance, C. E. RUSSET, *The Concept of Equilibrium in American Social Thought* (New Haven, Conn.: Yale University Press, 1966).
19. S. J. GOULD, "The Belt of an Asteroid," *Natural History,* Vol. 89, No. 1 (1980), pp. 26–33.

Chapter 7

1. A.N. WHITEHEAD, *Science and the Modern World,* p. 186.
2. *The Philosophy of Rudolf Carnap,* ed. P.A. SCHILPP (Cambridge: Cambridge University Press, 1963).

3. J. FRASER, "The Principle of Temporal Levels: A Framework for the Dialogue?" communication at the conference *Scientific Concepts of Time in Humanistic and Social Perspectives* (Bellagio, July 1981).
4. See on this point S. BRUSH, *Statistical Physics and Irreversible Processes,* esp. pp. 616–25.
5. Feuer has rather convincingly shown how the cultural context of Bohr's youth could have helped his decision to look for a non-mechanistic model of the atom; *Einstein and the Generation of Science* (New York: Basic Books, 1974). See also W. HEISENBERG, *Physics and Beyond* (New York: Harper & Row, 1971) and D. SERWER, "Unmechanischer Zwang: Pauli, Heisenberg and the Rejection of the Mechanical Atom 1923–1925," *Historical Studies in the Physical Sciences,* Vol. 8 (1977), pp. 189–256.
6. In *Black-Body Theory and the Quantum Discontinuity, 1894–1912* (Oxford: Clarendon Press and New York: Oxford University Press, 1978), Thomas Kuhn has beautifully shown how closely Planck followed Boltzmann's statistical treatment of irreversibility in his own work.
7. J. MEHRA and H. RECHENBERG, *The Historical Development of Quantum Theory,* Vols. 1–4 (New York: Springer Verlag, 1982).
8. See, about the conceptual framework of the experimental tests recently conceived for hidden variables in quantum mechanics, B. D'ESPAGNAT, *Conceptual Foundations of Quantum Mechanics,* 2nd aug. ed. (Reading, Mass.: Benjamin, 1976). See also B. D'ESPAGNAT, "The Quantum Theory and Reality," *Scientific American,* Vol. 241 (1979), pp. 128–40.
9. See, for the complementarity principle, B. D'ESPAGNAT, op. cit.; M. JAMMER, *The Philosophy of Quantum Mechanics* (New York: John Wiley & Sons, 1974); and A. PETERSEN, *Quantum Mechanics and the Philosophical Tradition* (Cambridge, Mass.: MIT Press, 1968). C. GEORGE and I. PRIGOGINE, "Coherence and Randomness in Quantum Theory," *Physica,* Vol. 99A (1979), pp. 369–82.
10. L. ROSENFELD, "The Measuring Process in Quantum Mechanics," *Supplement of the Progress of Theoretical Physics* (1965), p. 222.
11. About the quantum mechanics paradoxes, which can truly be said to be the nightmares of the classical mind, since they all (Schrödinger's cat, Wigner's friend, multiple universes) call to life again the phoenix idea of a closed objective theory (this time in the guise of Schrödinger's equation), see the books by d'Espagnat and Jammer.
12. B. MISRA, I. PRIGOGINE, and M. COURBAGE, "Lyapounov Vari-

able; Entropy and Measurement in Quantum Mechanics," *Proceedings of the National Academy of Sciences,* Vol. 76 (1979), pp. 4768–4772. I. PRIGOGINE and C. GEORGE, "The Second Law as a Selection Principle: The Microscopic Theory of Dissipative Processes in Quantum Systems," to appear in *Proceedings of the National Academy of Sciences.* Vol 80 (1983) 4590–94.
13. H. MINKOWSKI, "Space and Time," *The Principles of Relativity* (New York: Dover Publications, 1923).
14. A. D. SAKHAROV, *Pisma Zh. Eksp. Teor. Fiz.,* Vol. 5, No. 23 (1967).

Chapter 8

1. G. N. LEWIS, "The Symmetry of Time in Physics," *Science,* Vol. 71 (1930), p. 570.
2. A. S. EDDINGTON, *The Nature of the Physical World* (New York: Macmillan, 1948), p. 74.
3. M. GARDNER, *The Ambidextrous Universe: Mirror Asymmetry and Time-Reversed Worlds* (New York: Charles Scribner's Sons, 1979), p. 243.
4. M. PLANCK, *Treatise on Thermodynamics* (New York: Dover Publications, 1945), p. 106.
5. Quote by K. DENBIGH, "How Subjective Is Entropy?" *Chemistry in Britain,* Vol. 17 (1981), pp. 168–85.
6. See, for instance, M. KAC, *Probability and Related Topics in Physical Sciences* (London: Interscience Publications, 1959).
7. J. W. GIBBS, *Elementary Principles in Statistical Mechanics* (New York: Dover Publications, 1960), Chap. XII.
8. For instance, S. Watanabe introduces a strong distinction between the world to be contemplated and the world upon which we, as active agents, work; he states there is no consistent way of speaking about entropy increase if it is not in connection with our actions on the world. However, all our physics is in fact about the world to be acted on, and Watanabe's distinction thus does not help to clarify the relation between "microscopic deterministic symmetry" and "macroscopic probabilistic asymmetry." The question is left without an answer. How can we meaningfully say that the sun is irreversibly burning? See S. WATANABE, "Time and the Probabilistic View of the World," *The Voices of Time,* ed. J. FRASER (New York: Braziller, 1966).
9. Maxwell's demon appears in J. C. MAXWELL, *Theory of Heat* (London: Longmans, 1871), Chap. XXII; see also E. DAUB,

"Maxwell's Demon" and P. HEIMANN, "Molecular Forces, Statistical Representation and Maxwell's Demon," both in *Studies in History and Philosophy of Science,* Vol. 1 (1970); this volume is entirely devoted to Maxwell.

10. L. BOLTZMANN, *Populäre Schriften,* new ed. (Braunschweig-Wiesbaden: Vieweg, 1979). As Elkana emphasizes in "Boltzmann's Scientific Research Program and Its Alternatives," *Interaction Between Science and Philosophy* (Atlantic, Highlands, N.J.: Humanities Press, 1974), the Darwinian idea of evolution is explicitly expressed mostly in Boltzmann's view about scientific knowledge—that is, in his defense of mechanistic models against energeticists. See, for instance, his 1886 lecture "The Second Law of Thermodynamics," *Theoretical Physics and Philosophical Problems,* ed. B. MCGUINNESS (Dordrecht, Netherlands: D. Reidel, 1974).
11. For a recent account see I. PRIGOGINE, *From Being to Becoming—Time and Complexity in the Physical Sciences* (San Francisco: W. H. Freeman & Company, 1980).
12. In his *Scientific Autobiography,* Planck describes his changing relationship with Boltzmann (who was first hostile to the phenomenological distinction introduced by Planck between reversible and irreversible processes). See also on this point Y. ELKANA, op. cit., and S. BRUSH, *Statistical Physics and Irreversible Processes,* pp. 640–51; for Einstein, op. cit., pp. 672–74; for Schrödinger, E. SCHRÖDINGER, *Science, Theory and Man* (New York: Dover Publications, 1957).
13. H. POINCARÉ, "La mécanique et l'expérience," *Revue de Métaphysique et de Morale,* Vol. 1 (1893), pp. 534–37. H. POINCARÉ, *Leçons de Thermodynamique,* ed. J. Blondin (1892; Paris: Hermann 1923).
14. See for a study of the controversies around Boltzmann's entropy, see on this point S. BRUSH, *The Kind of Motion We Call Heat,* op. cit., and Planck's remarks in his biography (Loschmidt was Planck's student).
15. I. PRIGOGINE, C. GEORGE, F. HENIN, and L. ROSENFELD, "A Unified Formulation of Dynamics and Thermodynamics," *Chemica Scripta,* Vol. 4 (1973), pp. 5–32.
16. D. PARK, *The Image of Eternity: Roots of Time in the Physical World* (Amherst, Mass.: University of Massachusetts Press, 1980).
17. See also on this point S. BRUSH, *The Kind of Motion We Call Heat*—Book I, *Physics and the Atomists;* Book II, *Statistical Physics and Irreversible Processes* (Amsterdam: North Holland Publishing Company, 1976), as well as his commented anthology,

Kinetic Theory: Vol. I, *The Nature of Gases and Heat;* Vol. II, *Irreversible Processes* (Oxford: Pergamon Press, 1965 and 1966).

18. J. W. GIBBS, *Elementary Principles in Statistical Mechanics* (New York: Dover Publications, 1960), Chap XII. For an historical account, see J. MEHRA, "Einstein and the Foundation of Statistical Mechanics, *Physica,* Vol. 79A, No. 5 (1974), p. 17.
19. Many Marxist nature philosophers seem to take inspiration from Engels (quoted by Lenin in his *Philosophic Notebooks*) when he wrote in *Anti-Dühring* (Moscow: Foreign Languages Publishing House, 1954), p. 167, "Motion is a contradiction: even simple mechanical change of a position can only come about through a body being at one and the same moment of time both in one place and in another place, being in one and the same place and also not in it. And the continuous and simultaneous solution of this contradiction is precisely what motion is."
20. L. BOLTZMANN, *Lectures on Gas Theory* (Berkeley: University of California Press, 1964), p. 446f, quoted in K. POPPER, *Unended Quest* (La Salle, Ill.: Open Court Publishing Company, 1976), p. 160.
21. POPPER, op. cit., p. 160.

Chapter 9

1. VOLTAIRE, *Dictionnaire Philosophique*. (Paris: Garnier, 1954.)
2. See note 2, Chapter VII.
3. K. POPPER, "The Arrow of Time," *Nature,* Vol. 177 (1956), p. 538.
4. See M. GARDNER, *The Ambidextrous Universe,* pp. 271–72.
5. A. EINSTEIN and W. RITZ, *Phys. Zsch.,* Vol. 10 (1909), p. 323.
6. H. POINCARÉ, *Les méthodes nouvelles de la mécanique céleste* (New York: Dover Publications, 1957); E. T. WHITTAKER, *A Treatise on the Analytical Dynamics of Particles and Rigid Bodies* (Cambridge: Cambridge University Press, 1965).
7. J. MOSER, *Stable and Random Motions in Dynamical Systems* (Princeton, N.J.: Princeton University Press, 1974).
8. For a general review, see J. LEBOWITZ and O. PENROSE, "Modern Ergodic Theory," *Physics Today* (Feb. 1973), pp. 23–29.
9. For a more detailed study, see R. BALESCU, *Equilibrium and Non-Equilibrium Statistical Mechanics* (New York: John Wiley & Sons, 1975).
10. V. ARNOLD and A. AVEZ, *Ergodic Problems of Classical Mechanics* (New York: Benjamin, 1968).

11. H. POINCARÉ, "Le Hasard," *Science et Méthode* (Paris: Flammarion, 1914), p. 65.
12. B. MISRA, I. PRIGOGINE and M. COURBAGE, "From Deterministic Dynamics to Probabilistic Descriptions," *Physica,* Vol. 98A (1979), pp. 1–26.
13. D. N. PARKS and N. J. THRIFT, *Times, Spaces and Places: A Chronogeographic Perspective* (New York: John Wiley & Sons, 1980).
14. M. COURBAGE and I. PRIGOGINE, "Intrinsic Randomness and Intrinsic Irreversibility in Classical Dynamical Systems," *Proceedings of the National Academy of Sciences, 80* (April 1983).
15. I. PRIGOGINE and C. GEORGE, "The Second Law as a Selection Principle: The Microscopic Theory of Dissipative Processes in Quantum Systems," *Proceedings of the National Academy of Sciences,* Vol. 80 (1983), pp. 4590–4594.
16. V. NABOKOV, *Look at the Harlequins!* (McGraw-Hill 1974).
17. J. NEEDHAM, "Science and Society in East and West," *The Grand Titration* (London: Allen & Unwin, 1969).
18. See for more details B. MISRA, I. PRIGOGINE and M. COURBAGE, "From deterministic Dynamics to probabilistic Description", *Physica* 98A (1979) 1-26.; B. MISRA and I. PRIGOGINE "Time, Probability and Dynamics", in *Long-time Prediction in Dynamics,* eds. C. W. Horton, L. E. Recihl and A. G. Szebehely, (New York, Wiley 1983).
19. I. PRIGOGINE, C. GEORGE, F. HENIN, and L. ROSENFELD, "A Unified Formulation of Dynamics and Thermodynamics," *Chemica Scripta,* Vol. 4 (1973), pp. 5–32.
20. M. COURBAGE "Intrinsic irreversibility of Kolmogorov dynamical systems," *Physica* 1983; B. Misra and I. Prigogine, *Letters in Mathematical Physics,* September 1983.

Conclusion

1. A. S. EDDINGTON, *The Nature of the Physical World* (New York: Macmillan, 1948).
2. L. LÉVY-BRUHL, *La Mentalité Primitif* (Paris: PUF, 1922).
3. G. MILLS, *Hamlet's Castle* (Austin: University of Texas Press, 1976).
4. R. TAGORE, "The Nature of Reality" (Calcutta: *Modern Review* XLIX, 1931), pp. 42–43.
5. D. S. KOTHARI, *Some Thoughts on Truth* (New Delhi: Anniversary Address, Indian National Science Academy, Bahadur Shah Zafar Marg, 1975), p. 5.

6. E. MEYERSON, *Identity and Reality* (New York: Dover Publications, 1962).
7. Described in H. BERGSON, *Mélanges* (Paris: PUF, 1972), pp. 1340–46.
8. *Correspondence, Albert Einstein–Michele Besso, 1903–1955* (Paris: Herman, 1972).
9. N. WIENER, *Cybernetics* (Cambridge, Mass.: M.I.T. Press and New York: John Wiley & Sons, 1961).
10. M. MERLEAU-PONTY, "Le philosophe et la sociologie," *Éloge de la Philosophie,* Collection Idées (Paris: Gallimard, 1960), pp. 136–37.
11. M. MERLEAU-PONTY, *Résumés de Cours 1952–1960* (Paris: Gallimard, 1968), p. 119.
12. P. VALÊRY, *Cahiers,* La Pleiade (Paris: Gallimard, 1973), p. 1303.
13. For what follows see also I. PRIGOGINE, I. STENGERS, and S. PAHAUT, "La dynamique de Leibniz à Lucrèce," *Critique* "Spécial Serres," Vol. 35 (Jan. 1979), pp. 34–55. Engl. trans.: *"Dynamics from Leibniz to Lucretius,"* Afterword to M. SERRES, *Hermes: Literature, Science, Philosophy* (Baltimore: Johns Hopkins Univ. Pr., 1982), pp. 137–55.
14. C. S. PEIRCE, *The Monist* Vol. 2 (1892), pp. 321–337.
15. A. N. WHITEHEAD, *Process and Reality,* pp. 240–41. On this subject, see I. LECLERC, *Whitehead's Metaphysics* (Bloomington: Indiana University Press, 1975).
16. *La naissance de la physique dans le texte de Lucrèce,* p. 139.
17. LUCRETIUS, *De Natura Rerum,* Book II. "Again, if all movement is always interconnected, the new arising from the old in a determinate order—if the atoms never swerve so as to originate some new movement that will snap the bonds of fate, the everlasting sequence of cause and effect—what is the source of the free will possessed by living things throughout the earth?"
18. M. SERRES, op. cit., p. 136.
19. M. SERRES, op. cit., p. 162; also pp. 85–86 and "Roumain et Faulkner traduisent l'Écriture," *La traduction* (Paris: Minuit, 1974).
20. S. MOSCOVICI, *Hommes domestiques et hommes sauvages,* pp. 297–98.
21. T. KUHN, *The Structure of Scientific Revolutions,* 2nd ed. incr. (Chicago: Chicago University Press, 1970).
22. See A. N. WHITEHEAD, *Process and Reality,* op. cit. and M. HEIDEGGER *Sein und Zeit* (Tübingen: Niemeyer 1977).
23. H. WEYL, *Philosophy of Mathematics and Natural Science* (Princeton, N.J.: Princeton University Press, 1949).
24. A. NEHER, "Vision du temps et de l'histoire dans la culture juive," *Les cultures et le temps* (Paris: Payot, 1975), p. 179.

INDEX

Note: Page numbers given in **boldface** indicate location of definitions or discussions of terms or concepts indexed here.